Practical
Sedimentology

Practical Sedimentology

D. W. Lewis

University of Canterbury
New Zealand

Hutchinson Ross Publishing Company

Stroudsburg, Pennsylvania

An Apteryx Book

Copyright ©1984 by **Hutchinson Ross Publishing Company**
Library of Congress Catalog Card Number: 82-25839
ISBN: 0-87933-443-6

86 85 84   1 2 3 4 5
Manufactured in the United States of America.

Library of Congress Cataloging in Publication Data
Lewis, D. W., 1937–
    Practical sedimentology.
    Includes bibliographies and index.
    1. Sedimentology.   I. Title.
QE471.L46   1983   552′.5   82-25839
ISBN 0-87933-443-6

Distributed worldwide by Van Nostrand Reinhold Company Inc.,
135 W. 50th Street, New York, NY 10020.

Contents

Map Varieties   205

Suggestions for Fieldwork   215

About the Author   229

Preface

This manual resembles a debris flow in comprising flotsam (not jetsam!) from courses in which I try to treat the most practiced and practical aspects of sedimentary geology. The approach is similar to R.L. Folk's *Petrology of Sedimentary Rocks* (1965-1974), but the scope is broader. Emphasis is on providing illustrations and tables useful for description and interpretation of the common sedimentary deposits, and on providing "cookbook recipes" for utilitarian analytical techniques that do not require expensive equipment or advanced mathematical knowledge. The intent is to complement standard textbooks and to supplement individual suites of lectures and laboratory exercises at all levels beyond first year. The brief text in most sections is designed firstly to introduce the subject matter to students—without excessive preemption of any teachers' development of it—and secondly to review important aspects of the subject matter for the practicing postgraduate geologist.

Coverage of all sediment types is not attempted, and these are admitted deficiencies in the treatment of some aspects of all sections. Only a couple of exercises are included, in the belief that teachers prefer to generate their own. With the expectation of future revised editions of the manual, I solicit suggestions for improvements; if additional exercises are desired by users, I would be happy to incorporate others and would be pleased to receive examples. Any specific contributions utilized will be acknowledged.

ACKNOWLEDGMENTS

Constructive comments on preliminary drafts of this manual have been gratefully received from many professional colleagues and students. In particular, Professors D. S. Gorsline, University of Southern California, and K. A. W. Crook, Australian National University, made general suggestions and encouraged the construction of this more formalized version. Within the Geology Department, University of Canterbury, Glen Coates provided parts of the sections on grain-size methods and Ted Montague contributed much of the section on well logging. Sybil Tye and Wendy Nuthall ably typed the text, tables, and figure captions—how each maintained a happy and helpful disposition after interpreting my scrawl and retyping innumerable revisions is a mystery. Lee Leonard drafted final copies of the illustrations. Bernice Pettinato, of Hutchinson Ross Publishing Company, put a great deal of time and effort into the final editing for consistency and organization.

D. W. LEWIS

Introduction

The role of the sedimentologist is to interpret the history of sedimentary deposits. The scope of the subject is broad and entails both the study of processes involved in the origin, transportation, deposition, and postdepositional modification of sediments and the study of the lateral and vertical distribution of sedimentary deposits in relation to modern and ancient geographic, tectonic, and climatic settings. This manual begins with a brief review of environments in which sedimentation occurs, provides selected examples of vertical facies "models" used as guidelines for interpreting paleoenvironments, and then outlines physical and chemical processes that are important in sedimentation.

The bulk of the manual treats the three fundamental attributes of sedimentary deposits: structure, texture, and composition. Structures are treated superficially because of the plethora of excellent books and articles that emphasize them; in contrast to texture and composition, they primarily require study in the field. Each of these three attributes is imparted by a particular combination of processes acting in a particular geological framework. Interpretation from each attribute provides information about different combinations of processes and different aspects of the framework. Relationships can be summarized as follows:

$$D = S + T + C$$

where D is the sedimentary deposit, S is structure, T is texture, and C is composition.

$$E + G \xrightarrow{mcbt} S$$
$$E + P + D \xrightarrow{mcbt} T$$
$$P + Te + Cl(+ E + D) \xrightarrow{mcbt} C$$
$$E = f(Te + Cl + G)$$

where m, c, and b are, respectively, the mechanical, chemical, and biological processes and energy level and t is the time factor. E is the specific depositional environment, G is geography, Cl is climate, P is provenance, Te is tectonics, and D is diagenesis.

Final sections of the manual introduce principles of well logging, varieties of maps produced to represent geological interpretations, and some considerations necessary in fieldwork.

GENERAL DESCRIPTION OF SEDIMENTS AND SEDIMENTARY ROCKS

Description is the fundamental first step for geological investigations and reports. Early development of sound descriptive procedures will save time and effort in the long run; it will help you to avoid having to return to the outcrop or your samples, to avoid some misinterpretation, and to avoid confusion because of inadequate data.

The following characteristics should be described for all sedimentary deposits: color, induration, structure (sedimentary and tectonic), texture, and composition.

Color is best described by reference to the standard *Rock Colour Chart* (Goddard et al., 1951). The chart is based on the Munsell system, which is a widely accepted standard for color-identification. Colors are expressed by a shorthand notation: *hue* (given by a number) of red, yellow, green, blue, or purple (given by a letter); *value* of lightness or darkness of the color; and *chroma* of the strength or saturation of the color. For example, 5R 3/4 means hue of 5 red, value of 3, chroma of 4. Figure 1 shows the organization of the chart. Whether the material was wet or dry should be stated; if prac-

1

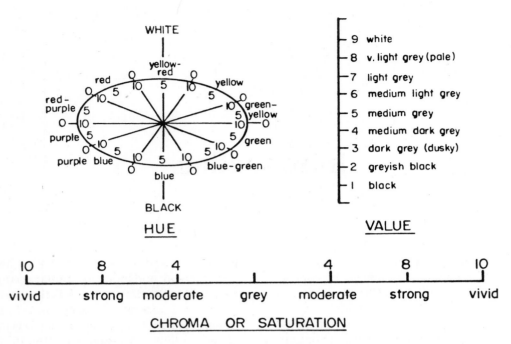

Figure 1. Organization of the standard rock color chart.

ticable, both wet and dry colors should be given. If without a chart, keep to primary colors—shades like "maroon," "chocolate brown," and "lilac," are visualized differently by different people.

Induration is somewhat subjective, but the following terms provide a useful standard: *loose; friable* (grains break free with slight finger pressure); *indurated; well indurated* (grains not broken when rock is broken); *very well indurated* (grains break when rock is broken)

Following chapters treat common sedimentary structures, the textures, and the composition of sediments.

LITERATURE

General Texts on Sedimentary Geology

Blatt, H., M. Middleton, and M. Murray, 1980, *Origin of Sedimentary Rocks*, 2nd ed., Prentice-Hall, Englewood Cliffs, N. J., 782p.

Dunbar, C. O., and J. Rogers, 1957, *Principles of Stratigraphy*, Wiley, New York, 356p.

Fairbridge, R. W., and J. Bourgeois, 1978, *The Encyclopedia of Sedimentology*, Dowden, Hutchinson & Ross, Stroudsburg, Pa., 928p.

Folk, R. L., 1980, *Petrology of Sedimentary Rocks*, Hemphill, Austin, Tex., 182p.

Friedman, G. M., and J. E. Sanders, 1978, *Principles of Sedimentology*, Wiley, New York, 792p.

Garrels, R. M., and F. T. MacKenzie, 1971, *Evolution of Sedimentary Rocks*, Norton, New York, 397p.

Goddard, E. N., D. D. Trask, R. K. de Ford, O. N. Rove, J. T. Singlewald, and R. M. Overbeck, 1951, *Rock Color Chart*, Geological Society of America, New York.

Grabau, A. W., 1913, *Principles of Stratigraphy*, 2 vols., Dover Publications, New York (1960 reprint), 1185p.

Greensmith, J. T., F. H. Hatch, and R. H. Rastall, 1971, *Petrology of the Sedimentary Rocks*, 5th ed., Thomas Murby, London, 502p.

Krumbein, W. C., and S. S. Sloss, 1963, *Stratigraphy and Sedimentation*, 2nd ed., Freeman, San Francisco, 660p.

Leeder, M. R., 1982, *Sedimentology*, Allen & Unwin, London, 344p.

Matthews, R. K., 1974, *Dynamic Stratigraphy*, Prentice-Hall, Englewood Cliffs, N. J., 370p.

Pettijohn, F. J., 1975, *Sedimentary Rocks*, 3rd ed., Harper & Row, New York, 628p.

Pettijohn, F. J., P. E. Potter, and R. Siever, 1972, *Sand and Sandstone*, Springer-Verlag, New York, 618p.

Reineck, H. -E., and I. B. Singh, 1980, *Depositional Sedimentary Environments*, 2nd ed., Springer-Verlag, New York, 549p.

Selley, R. C., 1976, *An Introduction to Sedimentology*, Academic Press, New York, 408p.

Twenhofel, W. A., et al., 1932, *Treatise on Sedimentation*, 2nd ed., Williams & Wilkins, Baltimore, 926p. (Reprinted in 1961 by Dover Publications, 2 vols., 926p.)

Twenhofel, W. A., 1950, *Principles of Sedimentation*, 2nd ed., McGraw-Hill, New York, 673p.

General References for Laboratory Techniques

Ali, S. H., and M. P. Weiss, 1968, Transmitted Infrared Photography: Cincinnatian Limestones, *Jour. Sed. Petrology* **38**:1350–1354.

Allman, M., and D. F. Lawrence, 1972, *Geological Laboratory Techniques*, Arco Publishing, New York, 335p.

Bouma, A. H., 1969, *Methods for the Study of Sedimentary Structures*, Wiley-Interscience, New York, 458p.

Carver, R. E., ed., 1971, *Procedures in Sedimentary Petrology*, Wiley-Interscience, New York, 653p. (Chapters by various authors on many aspects of sediment analysis).

Gilbert, G. M., and F. J. Turner, 1949, Use of the Universal Stage in Sedimentary Petrography, *Am. Jour. Sci.* **247**:1–26.

Griffiths, J. C., 1967, *Scientific Method in Analysis of Sediments*, McGraw-Hill, New York, 508p.

Jackson, M. L., 1958, *Soil Chemical Analysis—Advanced Course*, Prentice-Hall, Englewood Cliffs, N. J., 498p.

Hutchison, C. S., 1974, *Laboratory Handbook of Petrographic Techniques*, Wiley-Interscience, London, 527p.

Krumbein, W. C., and F. J. Pettijohn, 1938, *Manual of Sedimentary Petrography*, Appleton-Century-Crofts, New York, 549p.

Long, J. V. P., and S. O. Agrell, 1965, Cathodoluminescence of Minerals in Thin Section, *Mineralog. Mag.* **34**:318–326. (For an example in sedimentary petrology, see Sippel, R. E., 1968, Sandstone Petrology, Evidence from Luminescence Petrography, *Jour. Sed. Petrology* **38**:530–554.)

Maxwell, J. A., 1968, *Rock and Mineral Analysis*, Wiley-Interscience, New York, 584p. (Mainly for geochemistry.)

Merriam, D. F., ed., 1976, *Quantitative Techniques for the Analysis of Sediments: An International Symposium*, Pergamon Press, Elmsford, N.Y., 174p.

Milner, H. B., 1962, *Sedimentary Petrography*, vol. 1: *Methods in Sedimentary Petrography*; vol. 2: *Principles and Applications*, Allen & Unwin, London, 643p. and 715p. (Vol. 2 is particularly useful for mineral characteristics in sediments.)

Mueller, G., 1967. *Methods in Sedimentary Petrology*, H. -U. Schmincke, trans., Hafner, New York, 283p.

Nixon, W. C., 1969, Scanning Electron Microscopy, *Contemporary Physics* **10**:71–96. (For applications in sedimentary petrology, see, for example, Krinsley, D. H. and J. C. Doornkamp, 1973, *Atlas of Quartz Sand Surface Textures*, Cambridge University Press, Cambridge, 91p.)

Smykatz-Kloss, W., 1974, *Differential Thermal Analysis: Application and Results in Mineralogy*, Springer-Verlag, New York, 185p.

Tickell, F. G., 1965, *The Techniques of Sedimentary Mineralogy*, Developments in Sedimentology No. 4, Elsevier, Amsterdam, 220p.

Twenhofel, W. H., and S. A. Tyler, 1941, *Methods of Study of Sediments*, McGraw-Hill, New York, 183p.

Environments of Sedimentation

Interpretation of paleoenvironment is the ultimate goal of most sedimentary geological studies. In the broadest sense, the sedimentary environment encompasses the entire complex of physical, chemical, and biological materials and processes involved from the time sediments are formed to the time they are examined. Thus tectonic setting, geography, climate, provenance, diagenetic conditions, and late-stage (post-lithification) weathering effects must be considered. In the narrow sense considered in this section, the environment of sedimentation is the more restricted setting in which sediments actually accumulate.

A reasonably comprehensive list of depositional settings is provided in Fig. 2; many of these are illustrated in Fig. 3. Table 1 summarizes significant aspects of the marine environment in which most sediments are deposited. Selected general literature is listed (at the end of the chapter) to provide initial introduction to the vast number of publications relevant to paleoenvironmental interpretation.

Interpretation of the depositional environment requires synthesis of all inferences that can be made from the three-dimensional geometry and stratigraphic relationships of the various lithological units represented, the specific internal sedimentary structures within each unit (see separate section), the texture of each unit (see separate sections), and the composition (primarily of nondetrital components) insofar as it reflects chemical conditions (see separate sections). Post-depositional (diagenetic) modification to the original texture and composition must be evaluated and "subtracted" from the existing rock characteristics (see separate discussion). Geometry reflects setting, processes, and the time factor—the interaction of geographic influences, nature and energy levels and episodicity of transportational/depositional mechanisms, and the history of transgression or regression, progradation or retrogradation. Vertical stratigraphic sequences are among the most powerful aids to paleoenvironmental interpretation when there are no substantial erosional intervals or disconformities in the succession. The reason for their importance is that in such successions the superposed lithologies can be taken to represent originally laterally contiguous facies (an expression of "Walther's Law"—see Middleton, 1973). For example, in a marine setting with progradation or regression, boundaries between different subenvironments tend to move seaward, and if there is net accumulation, the result is a vertical stacking of lithological units with characteristics of progressively nearer-shore deposits. The reverse occurs with transgression (retrogradation implies erosion, and depositional sequences are less likely). Similar progressive overlapping of subenvironments occurs in nonmarine settings or without necessary relation to shorelines—for example, meandering river channels migrate across valley floors, and with net aggradation multistory sandy or gravelly deposits come to overlie finer levee and overbank deposits; similarly, lobes of submarine fans extend in various directions over the sea floor. Hence, a geologist can use the characteristics of a vertical succession of strata to reconstruct the original broad depositional setting without having to trace units out laterally or observing actual interfingering relationships. An exercise on transgression and

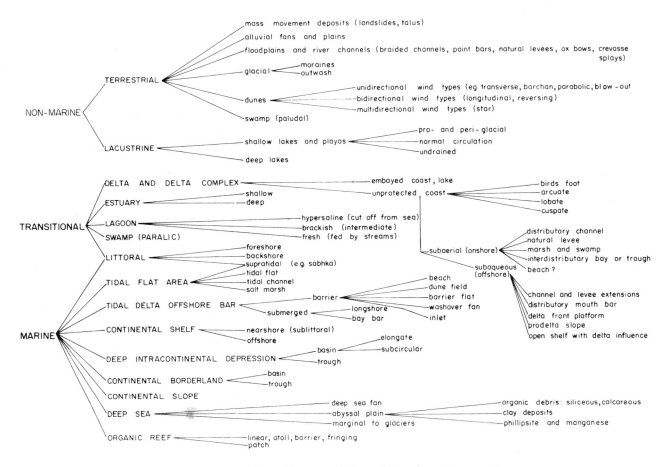

Figure 2. A classification of depositional environments.

regression at the end of this section illustrates applications of these concepts.

Sedimentary "models" or "facies models" have been devised for many environments. General or "ideal" models illustrate the "expected" associations and relationships; these are used as bases of comparison for the actual associations and relationships that occur in real sequences, and help guide interpretations. Specific models are constructed to explain the observed sequences (see Fig. 4); these may differ slightly or substantially from the ideal, and the recognition of differences leads the worker to consider explanations necessary to account for them. Feedback results in refinement of the general model. Examples of physical and stratigraphic models representing specific environments are given in Figs. 5–14 and Table 2.

Remember that geological interpretation is an exercise in compiling all relevant data, that it produces explanations that are most probable given the set of available data, and that in many cases definitive conclusions may not be possible—several possible explanations should be provided if the data are equivocal.

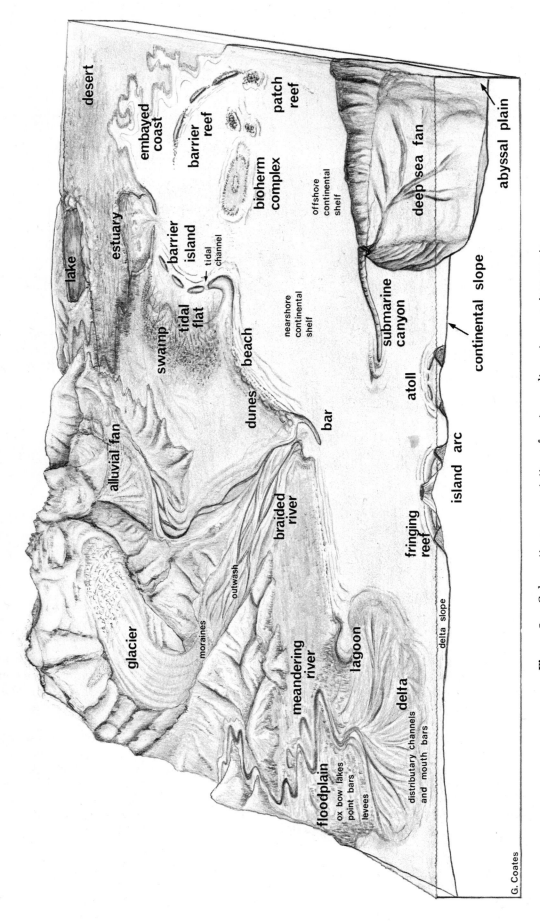

Figure 3. Schematic representation of major sedimentary environments. *Note:* Unrealistic environmental contractions are necessitated by representation of so many settings in one figure. (*Sketch by G. Coates*)

G. Coates

Table 1. Characteristics of the Marine Environment

Depth Zone:		Littoral	Shelf or Neritic (0–200 m)	Bathyal (200–1000 m)	Abyssal (1000+ m)
Inorganic Factors	Light	may be restricted by suspended matter; in clear tropical waters, intensity falls to 35% at 20m, 20% at 40m, 13% at 60m, 5% at 100m; different wavelengths with penetration depths		surface waters as neritic / Nil on substrate	
	Temperature	extreme variation	varies with latitude, weather, surficial currents	thermocline—rapid decrease with depth	stenothermal ca. 4°C
	Salinity	extreme variation	varies with latitude, weather, surficial currents	stenohaline	
	Turbulence	high or low	variable, storm waves effective	low; rare turbidity currents	
	Textures	gravels, sands; muds on some tidal flats	variable-sand, mud; well to poorly sorted (storm reworked or bioturbated)	muds with sandy turbidites; gravelly mass flow deposits near sub. canyons / calcareous ooze	siliceous ooze
	Structures	beach lamination, dune cross-bedding, flaser/lenticular bdg, ripple mark, mud cracks	variety-planar, cross and graded bdg, bioturbation	regular planar bdg, graded bdg; chaotic/contorted slide masses at base of slope	
	Authigenic/Perigenic components	beachrock cement; . . . . evaporites . . . .; broken shells . . . .; dolomite	glauconite; . . . . phosphates . . . .; . . . . pyrite . . . .; oolites	. . . . Mn nodules . . . .; sparse rhombs	
Benthic Biota	Flora	mangroves; . . . . sea grasses . . . .; blue green algae (mats and stromatolites); rhodoliths; calcified chlorophytes . . . . .; . . . . boring algae and fungi			
	Fauna	barnacles, bivalves, corals, foraminifera, bryozoa, cephalopods, echinoderms, annelids, crustacea (representatives in virtually all depth zones)			
	Traces	tracks . . . .; . . . . dwelling traces . . . .; . . . . feeding traces . . . .; . . . . grazing traces . . . .; (maximum diversity)			

7

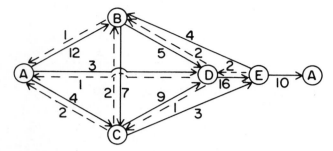

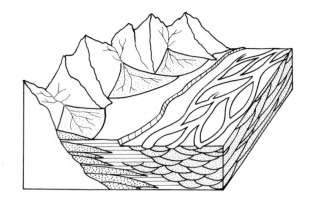

Figure 4. Method for determining facies model from stratigraphic interrelationships.

This is a facies relationship diagram showing numbers of transitions between lithotypes (letters) in a hypothetical formation. Lithotypes may be distinguished by texture and/or internal structures and/or composition. The characteristic succession can be determined by subjective evaluation of the raw numerical values, or by objective probability analysis of the numbers (see example in Walker, 1979, Chapter 1). Use of the diagram assumes Walthers Law and that no substantial unconformity is present. Diagrams may be improved by using different kinds of lines to indicate abrupt vs gradational transitions, or other features the worker considers important. This approach is particularly applicable to analysis of well logs and graphic logs (see later discussions).

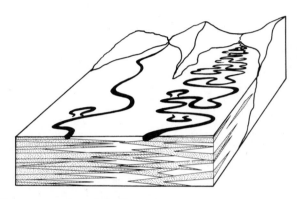

Figure 5. Simplified models illustrating settings and geometrical attributes of common alluvial facies. *Upper portion:* alluvial fans in piedmont setting and braided streams. *Lower portion:* low and high sinuosity meandering streams. *(After Allen, 1965)*

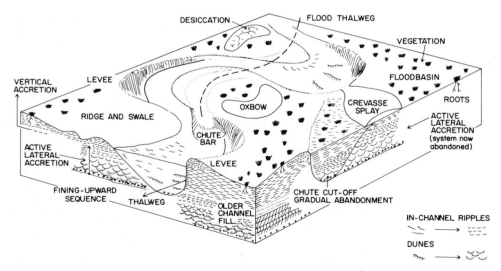

Figure 6. Meandering stream models. Major morphological elements of a meandering river system. Lateral accretion from point bar growth (convex bank) produces fining-upwards sequence, capped by vertical accretion deposits from overbank flooding. *(After R. G. Walker and Cant in Walker, 1979)*

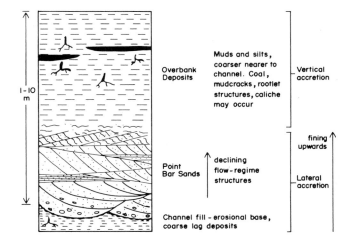

Overbank Deposits — Muds and silts, coarser nearer to channel. Coal, mudcracks, rootlet structures, caliche may occur — Vertical accretion

Point Bar Sands — declining flow-regime structures — Lateral accretion

fining upwards

Channel fill — erosional base, coarse lag deposits

1–10 m

Figure 6. *(Continued)*
Idealized fining-upwards sequence of aggraded meandering stream deposits. Complications may occur, for example, with sudden channel abandonment to form crevasse-splay deposits (wedges of coarse sediment fed through flood-breached channel levees—vertical profiles of these may resemble turbidite sequences) and/or ox-bow lakes (abandoned channel meander loops). Or, eolian sands may occur in the overbank succession.

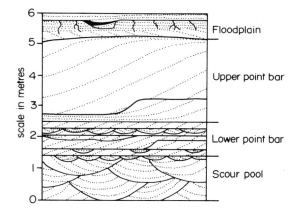

Floodplain

Upper point bar

Lower point bar

Scour pool

scale in metres

Figure 6. *(Continued)*
Cross sections of coarse-grained point bar to channel system. *(After McGowran and Garner, 1970)*

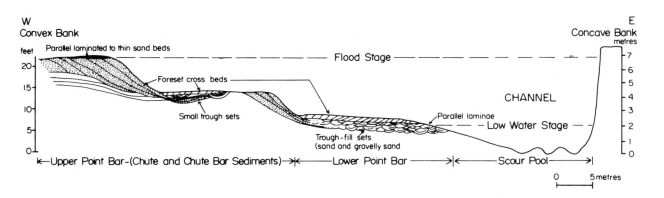

W
Convex Bank

feet

Parallel laminated to thin sand beds

Flood Stage

E
Concave Bank
metres

Foreset cross beds

Small trough sets

CHANNEL

Parallel laminae

Trough-fill sets (sand and gravelly sand)

Low Water Stage

|←—Upper Point Bar-(Chute and Chute Bar Sediments)—→|←———Lower Point Bar———→|←———Scour Pool———→|

0   5 metres

9

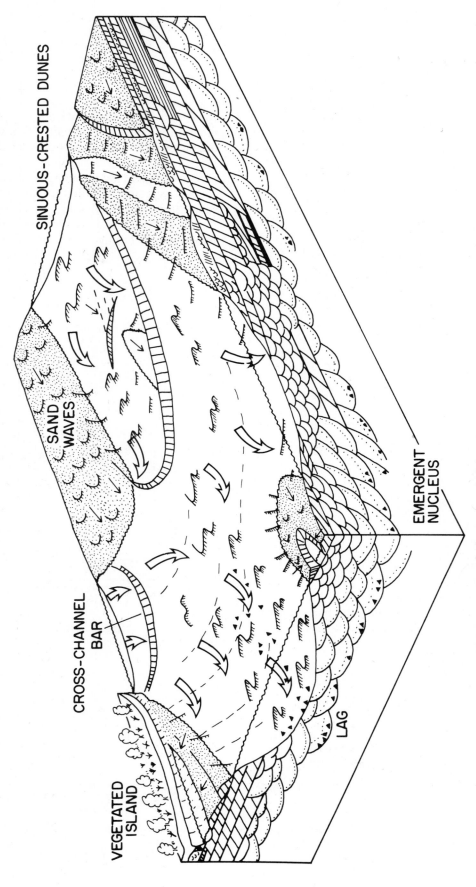

Figure 7. Braided stream models. (Figure continues on pages **11** and **12.** See also Table 2 on page **13.**) Major morphological elements of a braided river system. A vertical fining-upwards sequence may develop by burial of in-channel deposits below bar-top deposits. (*After R. G. Walker and D. J. Cant in R. G. Walker, 1979; vertical columns from Miall, 1978*)

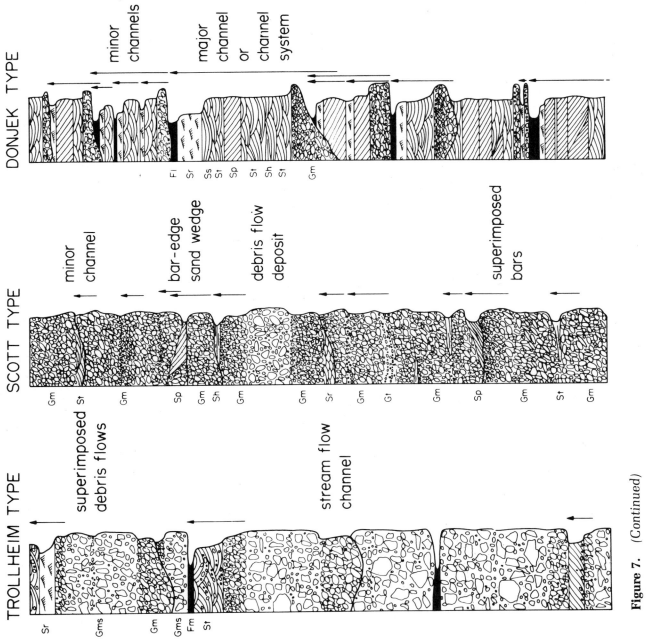

Figure 7. *(Continued)*

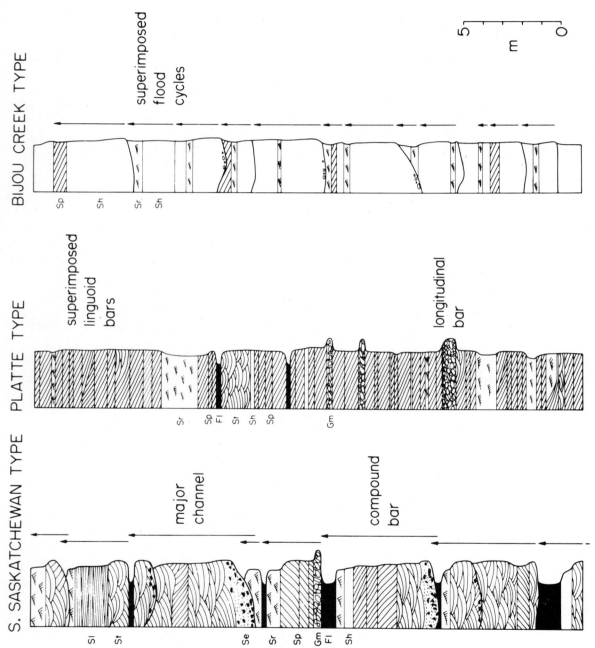

Figure 7. (Continued)

Table 2. Lithofacies and Interpretation for Six Principal Facies Assemblages
of Braided Stream Deposits

Trolheim type:	proximal rivers; predominantly alluvial fans subject to debris flows. Main facies: *Gms, Gm.* Also: *St, Sp, Fl, Fm.*
Scott type:	proximal rivers; predominantly stream flows and including alluvial fans. Main facies: *Gm.* Also: *Gp, Gt, Sp, St, Sr, Fl, Fm.*
Donjek type:	distal gravelly rivers; cyclic deposits. Main facies: *Gm, Gt, St.* Also: *Gp, Sh, Sr, Sp, Fl, Fm.*
South Saskatchewan type:	sandy braided rivers; cyclic deposits. Main facies: *St.* Also: *Sp, Se, Sr, Sh, Ss, Sl, Gm, Fl, Fm.*
Platte type:	sandy braided rivers; virtually noncyclic. Main facies: *St, Sp.* Also: *Sh, Sr, Ss, Gm, Fl, Fm.*
Bijou Creek type:	ephemeral or perennial rivers subject to flash floods. Main facies: *Sh, Sl.* Also: *Sp. Sr.*

Facies Code	Lithofacies	Sedimentary Structures	Interpretation
Gms	massive, matrix-supported gravel	none	debris flow deposits
Gm	massive or crudely bedded gravel	horizontal bedding, imbrication	longitudinal bars, lag deposits, sieve deposits
Gt	gravel, stratified	trough crossbeds	minor channel fills
Gp	gravel, stratified	planar crossbeds	linguoid bars or deltaic growths from older bar remnants
St	sand, medium to very coarse, may be pebbly	solitary (theta) or grouped (pi) trough crossbeds	dunes (lower flow regime)
Sp	sand, medium to very coarse, may be pebbly	solitary (alpha) or grouped (omikron) planar crossbeds	linguoid, transverse bars, sand waves (lower flow regime)
Sr	sand, very fine to coarse	ripple marks of all types	ripples (lower flow regime)
Sh	sand, very fine to very coarse, may be pebbly	horizontal lamination, parting or streaming lineation	planar bed flow (lower and upper flow regime)
Sl	sand, fine	low angle ($<10°$) crossbeds	scour fills, crevasse splays, antidunes
Se	erosional scours with intraclasts	crude crossbedding	scour fills
Ss	sand, fine to coarse, may be pebbly	broad, shallow scours including eta cross-stratification	scour fills
Sse, She, Spe	sand	analogous to *Ss, Sh, Sp*	eolian deposits
Fl	sand, silt, mud	fine lamination, very small ripples	overbank or waning flood deposits
Fsc	silt, mud	laminated to massive	backswamp deposits
Fcf	mud	massive, with freshwater molluscs	backswamp pond deposit
Fm	mud, silt	massive, desiccation cracks	overbank or drape deposits
Fr	silt, mud	rootlets	seatearth
C	coal, carbonaceous mud	plants, mud films	swamp deposits
P	carbonate	pedogenic features	soil

Source: after Miall, 1977

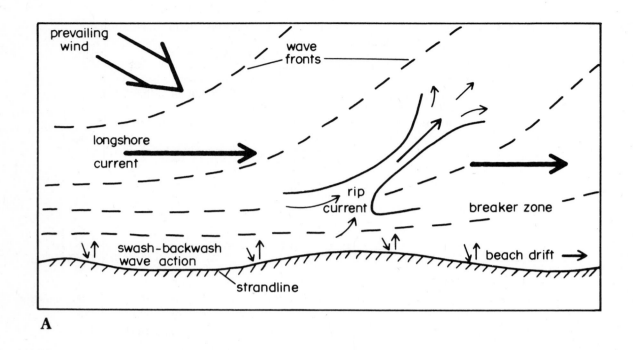

A

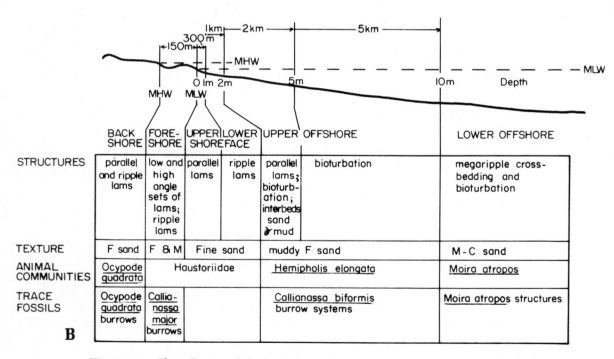

	BACK SHORE	FORE-SHORE	UPPER SHOREFACE	LOWER SHOREFACE	UPPER OFFSHORE		LOWER OFFSHORE
STRUCTURES	parallel and ripple lams	low and high angle sets of lams; ripple lams	parallel lams	ripple lams	parallel lams; bioturb-ation; interbeds sand & mud	bioturbation	megaripple cross-bedding and bioturbation
TEXTURE	F sand	F & M	Fine sand		muddy F sand		M - C sand
ANIMAL COMMUNITIES	Ocypode quadrata	Haustoriidae			Hemipholis elongata		Moira atropos
TRACE FOSSILS	Ocypode quadrata burrows	Callia-nassa major burrows			Callianassa biformis burrow systems		Moira atropos structures

B

Figure 8.   Shoreline models. *A*: Schematic of nearshore system of sediment dynamics. *B*: Sample of facies in beach and nearshore environment, Sapelo Island, Georgia. *(B only after Howard and Reineck, 1972)*

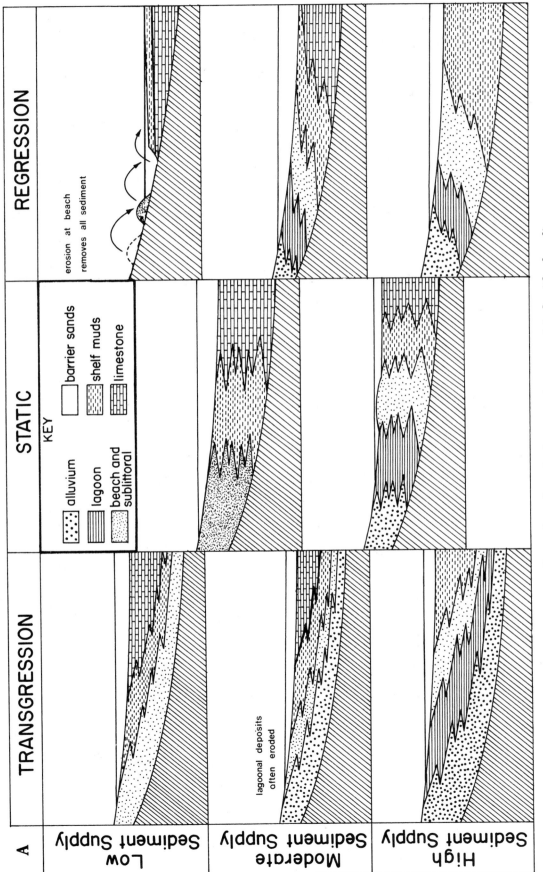

Figure 9. Examples of vertical successions associated with shorelines. *(Figure continues on next page.)* A: Idealized relationships between sedimentary facies in linear shorelines supplied with detrital sediments. *(After Selley, 1970)*

KEY

- alluvium
- lagoon
- beach and sublittoral
- barrier sands
- shelf muds
- limestone

TRANSGRESSION — STATIC — REGRESSION

Low Sediment Supply

Moderate Sediment Supply

High Sediment Supply

lagoonal deposits often eroded

erosion at beach removes all sediment

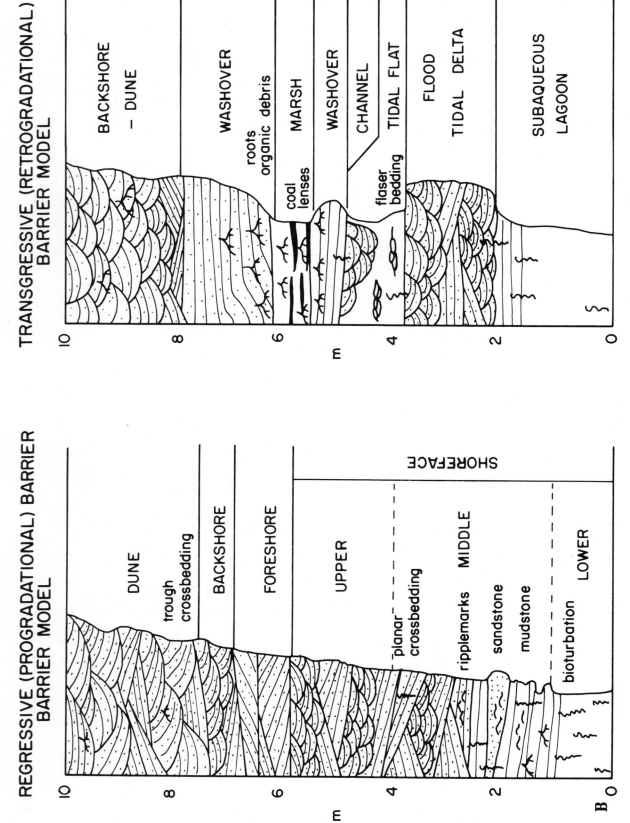

REGRESSIVE (PROGRADATIONAL) BARRIER
BARRIER MODEL

DUNE

trough
crossbedding

BACKSHORE

FORESHORE

SHOREFACE

UPPER

planar
crossbedding

ripplemarks   MIDDLE

sandstone

mudstone

bioturbation   LOWER

B

TRANSGRESSIVE (RETROGRADATIONAL)
BARRIER MODEL

BACKSHORE
– DUNE

WASHOVER

roots
organic debris

MARSH

WASHOVER

CHANNEL

TIDAL FLAT

FLOOD

TIDAL DELTA

SUBAQUEOUS
LAGOON

coal
lenses

flaser
bedding

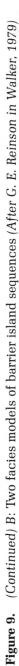

Figure 9. *(Continued) B: Two facies models of barrier island sequences (After G. E. Reinson in Walker, 1979)*

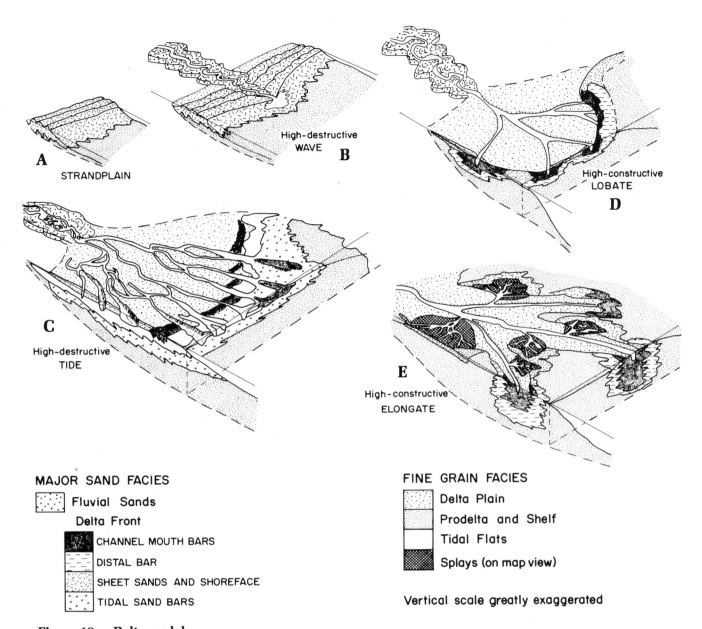

MAJOR SAND FACIES

▫ Fluvial Sands

Delta Front

■ CHANNEL MOUTH BARS

▫ DISTAL BAR

▫ SHEET SANDS AND SHOREFACE

▫ TIDAL SAND BARS

FINE GRAIN FACIES

▫ Delta Plain

▫ Prodelta and Shelf

▫ Tidal Flats

▫ Splays (on map view)

Vertical scale greatly exaggerated

Figure 10. Delta models.

Framework facies in major types of deltas. A: *Strand-plain,* shown for comparison. B: *High-destructive wave* delta, composed primarily of shoreface and associated fluvial sands. C: *High-destructive tidal* delta, with extensive tidal shoal or sandflat facies.

D: *High-constructive lobate* delta, with associated fluvial sands, channel mouth bars and delta front sheet sands. E: *High-constructive elongate* delta, with thick channel mouth bars or bar fingers. *(After A. J. Scott in Fisher et al., 1969)*

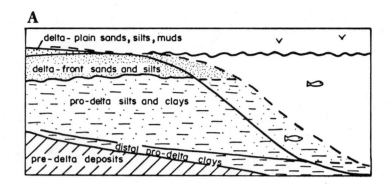

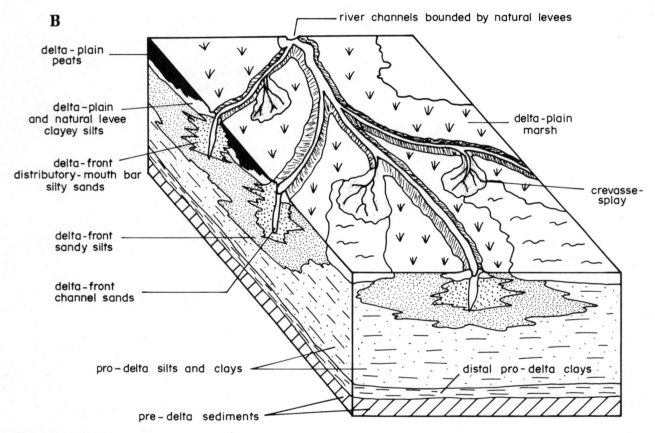

Figure 11. Delta models.

A: Idealized longitudinal cross-section through high-constructive delta. Dashed lines represent configuration of deltaic deposits after further progradation.

B: Idealized model of high-constructive delta, showing paleoenvironments and resultant facies.

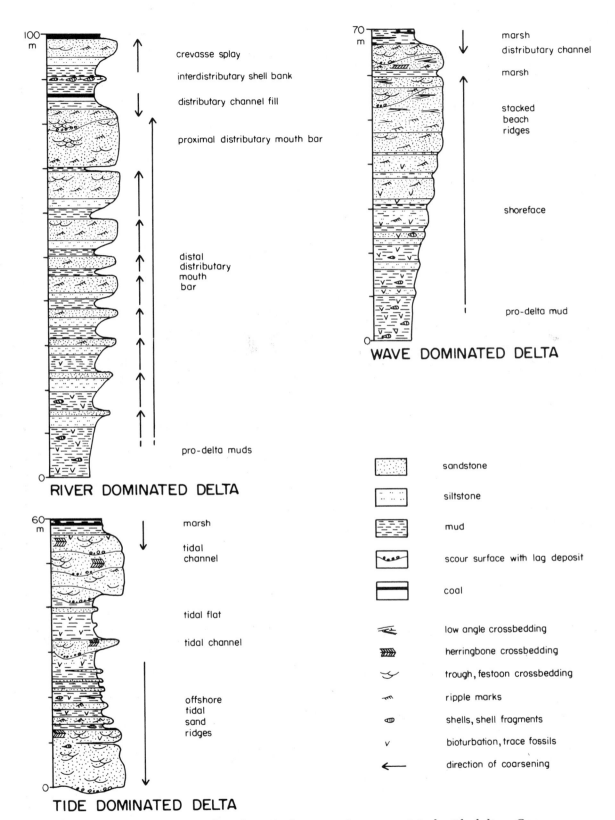

Figure 12. Examples of vertical successions associated with deltas. Generalized representative vertical sections for three principal varieties of deltas. *(After A. D. Miall in Walker, 1979)*

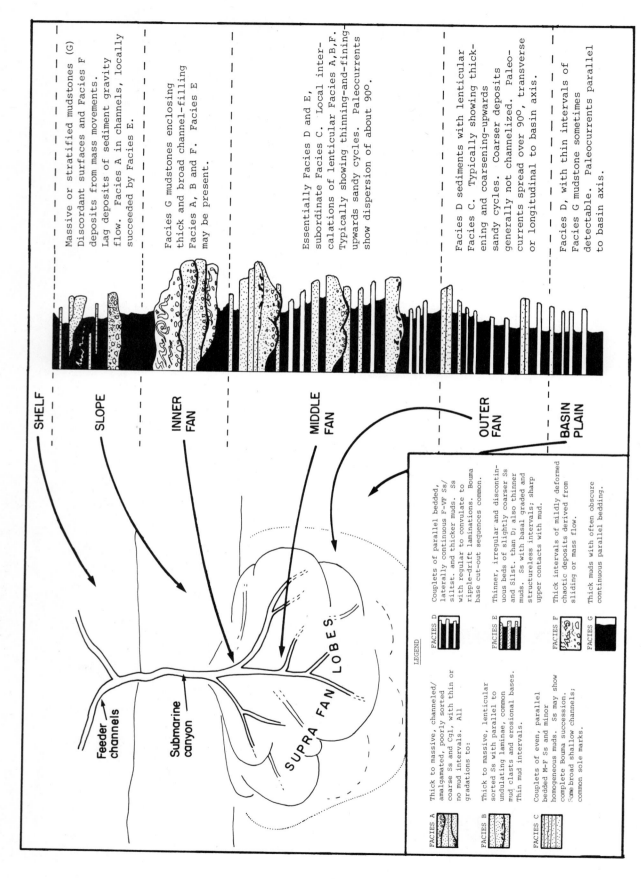

Massive or stratified mudstones (G) Discordant surfaces and Facies F deposits from mass movements. Lag deposits of sediment gravity flow. Facies A in channels, locally succeeded by Facies E.

Facies G mudstones enclosing thick and broad channel-filling Facies A, B and F. Facies E may be present.

Essentially Facies D and E, subordinate Facies C. Local intercalations of lenticular Facies A,B,F. Typically showing thinning-and-fining upwards sandy cycles. Paleocurrents show dispersion of about 90°.

Facies D sediments with lenticular Facies C. Typically showing thickening and coarsening-upwards sandy cycles. Coarser deposits generally not channelized. Paleocurrents spread over 90°, transverse or longitudinal to basin axis.

Facies D, with thin intervals of Facies G mudstone sometimes detectable. Paleocurrents parallel to basin axis.

SHELF

SLOPE

INNER FAN

MIDDLE FAN

OUTER FAN

BASIN PLAIN

Feeder channels

Submarine canyon

SUPRA FAN LOBES

LEGEND

FACIES A   Thick to massive, channeled/amalgamated, poorly sorted coarse Ss and Cgl. with thin or no mud intervals. All gradations to:

FACIES B   Thick to massive, lenticular sorted Ss with parallel to undulating laminae, common mud clasts and erosional bases. Thin mud intervals.

FACIES C   Couplets of even, parallel bedded M-F Ss and minor homogeneous muds. Ss may show complete Bouma succession. Some broad shallow channels; common sole marks.

FACIES D   Couplets of parallel bedded, laterally continuous F-VF Ss/siltst. and thicker muds. Ss with regular to convulate to ripple-drift laminations. Bouma base cut-out sequences common.

FACIES E   Thinner, irregular and discontinuous beds of slightly coarser Ss and Silst. than D; also thinner muds. Ss with basal graded and structureless intervals; sharp upper contacts with mud.

FACIES F   Thick intervals of mildly deformed chaotic deposits derived from sliding or mass flow.

FACIES G   Thick muds with often obscure continuous parallel bedding.

Figure 13.   Submarine Fan Model. (Column after Mutti and Ricchi-Lucchi, 1978)

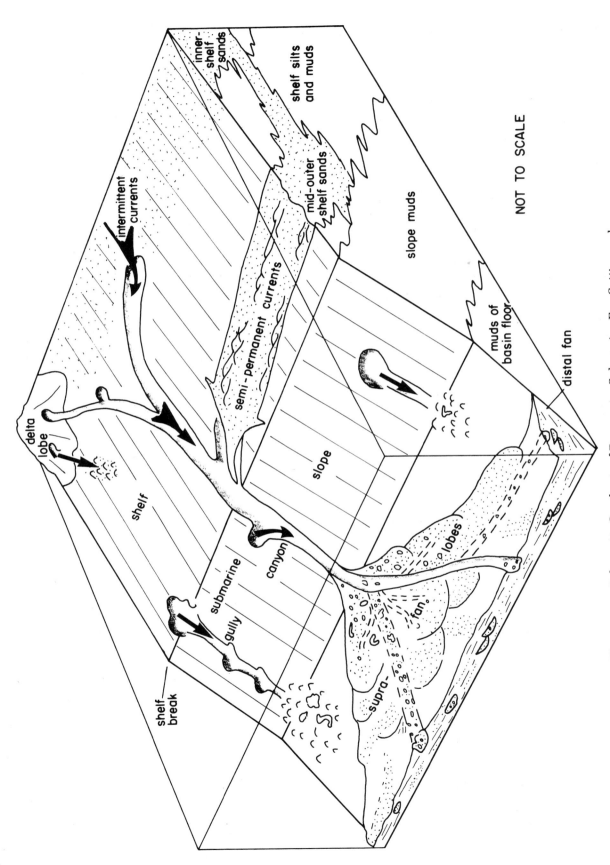

NOT TO SCALE

Labels within figure:
- inner-shelf sands
- shelf silts and muds
- intermittent currents
- mid-outer shelf sands
- slope muds
- semi-permanent currents
- muds of basin floor
- distal fan
- delta lobe
- shelf
- slope
- submarine canyon
- lobes
- gully
- supra-fan
- shelf break

Figure 14. Schematic Synthesis of Deepwater Submarine Fan Setting and Sediment Supply System. (Reproduced from Lewis, 1982, New Zealand Journal of Geology and Geophysics)

AN EXERCISE ON TRANSGRESSION-REGRESSION AND THE DEVELOPMENT OF SEDIMENTARY ASSOCIATIONS

1. The accompanying map (Fig. 15) indicates sediment types in an idealized nonmarine to offshore marine environment. Along the lines of section given (AA' and BB'), construct cross-sections showing the vertical succession of lithotypes that would result from the series of events listed below. Assume that rates of deposition are the same for each sediment type, that the vertical thickness of each lithotype is independent of the map width of its depositional environment, and that boundaries between lithotypes maintain (unrealistically) the same relative position. For the vertical scale, use 1 cm = 1 m. Computer chart paper is ideal for this exercise (use appropriate vertical scale).

(a) Regression for 7 km during which 2 m of sediment accumulates.

(b) Transgression for 5 km during which 3 m of sediment accumulates.

(c) Transgression for 4 km during which 2 m of sediment accumulates.

(d) Regression for 6 km during which 1 m of sediment accumulates.

(e) Hold for accumulation of 3 m of sediment.

(f) Transgression for 10 km during which 3 m of sediment accumulates.

2. Draw a cross-section C-C'.

3. Draw a cross-section of B-B' as if there were an erosional episode of 2 m uniformly over the region after event D.

4. Draw a cross-section A-A' as if there were:

(a) continuous (2 km/event) progradation of the delta.

(b) continuous (1 km/event) retrogradation of the delta.

5. In geological investigations you might be restricted to constructing only one cross-section from field or borehold data. Would you reach similar conclusions concerning the history of the succession from cross-section A-A' as from C-C'? Could you identify transgressions/regressions and the direction of land from both?

6. When sufficient field or borehole data are available, a "panel" or "fence" diagram can be drawn to represent the three-dimensional stratigraphy (see chapter on map varieties). Sketch such a diagram, using points 1, 2, 3, 4, 5. Is there any plan-view map that would show the relationships reasonably well?

7. In realistic examples, is the rate of either transgression or regression likely to be indicated by the thickness of the lithotypes?

8. Draw some "time planes" on your cross-sections. Formulate a general statement showing the relationship of time planes and rock units in (a) transgressive and (b) regressive sequences.

9. Why might faunal studies not provide time planes as definitive as those you have marked? Would you have been able to identify any isochronous planes had you observed rather than constructed the stratigraphic sections?

10. How would you design a program to collect samples (such as fauna) that differed sequentially in time throughout the span of deposition of the silt lithotype? What limitations are there to biostratigraphic data collected by paleontologists from vertical sections in the field or from boreholes?

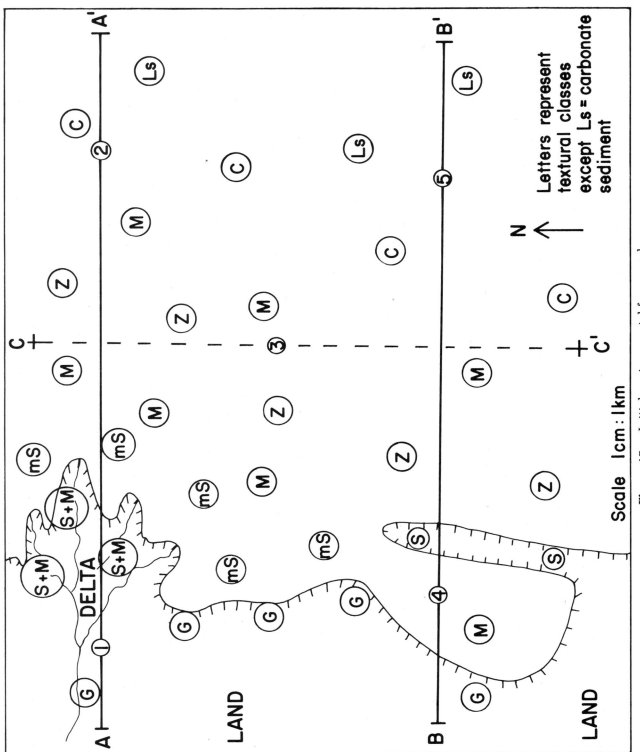

Figure 15. Initial environmental framework.

Scale 1cm : 1km

Letters represent textural classes except Ls = carbonate sediment

23

LITERATURE

Allen, J. R. L., 1965, A review of the origin and characteristics of recent alluvial sediments, *Sedimentology* 5:89-191.

Bouma, A. H., 1962, *Sedimentology of Some Flysch Deposits*, Elsevier, Amsterdam, chap. 1 (vertical profile technique).

Conybeare, C. E. B., 1979, *Lithostratigraphic Analysis of Sedimentary Basins*, Academic Press, New York, 555p.

Cook, H. E., and P. Enos (eds.), 1977, *Deep-Water Carbonate Environments*, SEPM Spec. Pub. No. 25, Society of Economic Paleontologists and Mineralogists, Tulsa, Okla., 336p.

Curtis, D. M. (ed.), 1978, *Depositional Environments and Paleoecology: Environmental Models in Ancient Sediments*, SEPM Rept. Series, No. 6, Society of Economic Paleontologists and Mineralogists, Tulsa, Okla., 240p.

Douglas, R. G., I. P. Colburn, and D. S. Gorsline, 1981, *Depositional Systems of Active Continental Margin Basins*, Short Course notes, SEPM Pacific Section, Society of Economic Paleontologists and Mineralogists, Tulsa, Okla., 165p.

Doyle, L. J., and O. H. Pilkey, Jr. (eds.), 1979, *Geology of Continental Slopes*, SEPM Spec. Pub. No. 27, Society of Economic Paleontologists and Mineralogists, Tulsa, Okla., 374p.

Ethridge, F. G., and R. M. Flores (eds.), 1981, *Recent and Ancient Nonmarine Depositional Environments: Models for Exploration*, SEPM Spec. Pub. No. 31, Society of Economic Paleontologists and Mineralogists, Tulsa, Okla., 349p.

Fisher, W. L., L. F. Brown, Jr., A. J. Scott, and J. H. McGowan, 1969, *Delta Systems in the Exploration for Oil and Gas*, Bureau of Economic Geology, University of Texas at Austin.

Friedman, G. M. (ed.), 1969, *Depositional Environments in Carbonate Rocks*, SEPM Spec. Pub. No. 14, Society of Economic Paleontologists and Mineralogists, Tulsa, Okla., 209p.

Howard, J. D., and H.-E. Reineck, 1972, Georgia Coastal Region, U.S.A.: Sedimentology and biology. VII: Physical and biogenic sedimentary structures of the nearshore shelf, *Senckenbergiana marit.* 7:217-222.

Lewis, D. W., 1982, Channels across continental shelves: Corequisites of canyon-fan systems and potential petroleum conduits, *New Zealand Jour. Geology and Geophysics* 25:209-255.

McGowran, J. H., and L. E. Garner, 1970, Physiographic features and stratification types of coarse-grained point bars: modern and ancient examples, *Sedimentology* 14:77-111.

Miall, A. D. (ed.), 1978, *Fluvial Sedimentology*, Canadian Soc. Petroleum Geologists Mem. 5, 859p.

Middleton, G. V., 1973, Johannes Walther's Law of the Correlation of Facies; *Geol. Soc. America Bull.* 84:979-988.

Morgan, J. P., and R. H. Shaver (eds.), 1970, *Deltaic Sedimentation Modern and Ancient*, SEPM Spec. Pub. No. 15, Society of Economic Paleontologists and Mineralogists, Tulsa, Okla., 312p.

Mutti, E., and F. Ricci-Lucci, 1978, Turbidites of the northern Appenines: introduction to facies analysis, *Internat. Geology Rev.* 20:125-166.

Reading, H. G. (ed.), 1978, *Sedimentary Environments and Facies*, Blackwell Scientific Publications, Boston, 557p.

Reineck, H.-E., and I. B. Singh, 1973 (2nd ed., 1980), *Depositional Sedimentary Environments*, 2nd ed., Springer-Verlag, New York, 439p.

Rigby, J. K., and W. K. Hamblin (eds.), 1972, *Recognition of Ancient Sedimentary Environments*, SEPM Spec. Pub. No. 16, Society of Economic Paleontologists and Mineralogists, Tulsa, Okla., 340p.

Scholle, P. A., and D. Spearing (eds.), 1982, *Sandstone Depositional Environments*, AAPG Memoir 31, American Association of Petroleum Geologists, Tulsa, Okla., 410p.

Selley, R. C., 1970, *Ancient Sedimentary Environments*, Cornell University Press, Ithaca, N.Y., 237p.

Shepard, F. P., 1964, Criteria in modern sediments useful in recognizing ancient sedimentary environments, in L.M.J.U. van Straaten, (ed.), *Deltaic and Shallow Marine Deposits*, Elsevier, Amsterdam, 1-25.

Spearing, D. R., 1974, Summary sheets of sedimentary deposits, *Geol. Soc. America Maps and Charts* **M-8**, 6 sheets.

Walker, R. G. (ed.), 1979, *Facies Models*, Geological Association of Canada reprint series no. 1, Geoscience Canada, Toronto, 211p.

Wolf, K. H., 1973, Conceptual models 1: examples in sedimentary petrology, environmental and stratigraphic reconstructions and soil, reef, chemical and placer sedimentary ore deposits, *Sed. Geology* 9:153-93. (Also see Wolf's Conceptual models 2, Fluvial-alluvial, glacial, locustrine, desert and shorezone [bar-beach-dune-chenier] milieus, Sed. Geology 9:261-281.)

Processes in Sedimentation

TRANSPORTATION

Whereas the ultimate goal of many studies of sedimentary deposits is the determination of paleoenvironment, it must be clearly realized that the observable characteristics of a specific depositional unit permit interpretation only of *process*, not directly of depositional environment. The paleoenvironment is inferred after synthesis of the suite of characteristics of a vertical (and, if feasible, a lateral) association of units, generally via analogy with a "model" association of facies determined from previous studies (for example, of modern environments). The same process may act in various environments. Some are very episodic (*large* storms) yet have a dominant effect on the character of sequences (for example, see Einsele and Seilacher, 1982). The following discussion introduces some considerations with respect to determining mechanical processes operating at the time of deposition, with an emphasis on gravity-flow processes, which most texts treat less fully.

In general, mechanical deposition may occur from any of the following:

Passive suspension: by gravitational settling through a medium that is not actively transporting particles (such as clays in a lake, plankton in the deep sea, ash from volcanic eruptions).

Entrainment flow: by direct deposition from air or water that has been actively transporting the particles.

Gravity flow: by direct deposition from particles that have been moving down a slope because of the gravitational acceleration acting on them. The term *resedimentation* has been widely used for gravity-flow processes, but unwisely, because most sediments are redeposited by a variety of processes many times before finally coming to rest in the setting we find them. Also, some sediments may be transported from their source to their final site of accumulation by a single episode of gravity flow.

Distinction among these classes of processes is not always clear cut; for example, air or water is commonly moving when suspended sediments settle and may have some transporting effect, and gravity flows may entrain some particles as a traction population. However, the characteristics of deposits commonly indicate which major class of process *dominated* during deposition.

Passive Suspension

Structures are either absent (massive units) or consist of diffuse or sharp planar stratification between units deposited from different episodes of sedimentation. Stratification generally forms perpendicular to the gravitational field, but mantles and conforms to irregularities present on the underlying surface. Layers may be very thin and regular (for example, varves).

Textures are generally fine (clay to fine sand size), but there are exceptions (such as pebbles falling from "rafts" of ice or floating tree roots). Normal grading from coarse up to fine is expected but may be difficult to determine if particle size is extremely fine (as with varves). For each particle, the plane that contains the long and the intermediate axis will be deposited perpendicular to the gravity field. (*Note:* postdepositional compaction may also impart a parallel fabric.)

Entrainment Flows

Particles entrained and transported by water or air currents move either as a traction load (continuous contact with the bed of the flow), a saltation load (intermittent contact with the bed) and/or an (active) suspension load (no contact with the bed). Most sediments are deposited from entrainment flows and these processes can be assumed if criteria for the other mechanisms are lacking.

Structures are characteristically ripplemarks, dune bedforms, cross-lamination, and cross-bedding, but these may not be present or may not be visible (such as where textures and composition are homogeneous). Bottom contacts commonly are erosion surfaces, but the erosion may reflect an episode not connected with deposition of the unit itself—hence an erosional base is not definitive. Regular or irregular, initially subhorizontal, well-defined stratification between units is common but not diagnostic.

Textures generally show some evidence of sorting, and most well-sorted sediments were deposited from entrainment flows; however, sorting is not a definitive criterion. Well-sorted deposits with imbrication of particles are diagnostic. Normal or inverse grading is possible. Well-developed particle fabric is common (imbrication of pebbles, parallel alignment of the long axes of sand grains, and so on), but it can develop in some gravity flows and may not be apparent in entrainment-flow deposits.

If deposition from entrainment flow is known or assumed, textural analysis together with sedimentary structures can provide additional information with respect to process (see chapters on sedimentary structures and on texture of detrital sediments; also Middleton, 1965; Allen, 1970; Moss, 1972; Stanley and Swift, 1976; Wright, 1977; Middleton and Southard, 1978).

Gravity Flows

Considerable work has been undertaken and is underway on sediment gravity flows and their deposits; concepts and characteristics are being constantly refined (for example, see Middleton and Hampton, 1973; Carter, 1975; Lewis, 1976, 1980; Lowe, 1981, 1982; Field, 1981; Pierson, 1981). Definitive structural and textural attributes are not yet certain because of the high variability found, although the presence of outsize clasts is suggestive of gravity-flow deposits (ice- or plant-rafted pebbles and animal gizzard stones are exceptions). Some sorted and homogeneous sandy gravity-flow deposits are very difficult to distinguish from entrainment-flow deposits.

Sediment gravity flows move because of gravitational acceleration acting on the particles—hence movement is essentially down the slope, and deposition occurs where the downslope component of that acceleration is less than the variety of retarding forces acting on the particles. Where resistance to flow is low, and depending on the energy attained by the flow down its initial slope, transport over very low gradients may be as great as several tens of kilometers in subaerial settings and several hundreds of kilometers (to more than 1000 km, according to some interpretations) in marine settings. Given the presence of a slope, an unstable accumulation of sediment, and a trigger mechanism, gravity-flow deposits may occur in any environment. However, three settings appear to be particularly common: alluvial fans, particularly in settings of high relief; submarine fans, generally in front of submarine canyons, and vertically stacked submarine channels in a shelf-type environment, landward of any submarine canyon (see Lewis, 1982; and Fig. 14). The following are criteria for distinction:

Alluvial fan: dominance of subangular to subround gravelly detritus; paucity of fine sediment, especially mud; stratification lensoidal on the scale of a substantial exposure; poor sorting; absence of marine biota; evidence of *in situ* pedogenetic weathering (particularly red iron oxides) or plant remains; local sorted (fluvial) lenses, becoming dominant distally; associated nonmarine facies.

Submarine fan: intercalation of conglomerates, sandstones, and mudstones, with conglomerates dominant proximally; bathyal to abyssal marine biota in intercalated mudstones; presence of continuous sandstone or mudstone beds on the scale of the outcrop; gravel fraction a mixture of detritus (commonly rounded but possibly angular) and perigenic clasts (derived from shoreward sediments that were not fully consolidated); most conglomerates with a muddy matrix; slump masses proximally. Surrounding marine facies generally dominated by mudstones with sandstones of "turbidite" aspect.

Shelf channel: broad, shallow channels incised into mudstones and fine sandstones and filled by conglomerates and coarser sandstones; marine biota of neritic to upper bathyal character in intercalated mudstones; presence of continuous units on the scale of the exposure; conglomerates commonly clast-supported with sandy or gravelly matrix, but muddy matrix-supported conglomerates may be present; surrounding sediments often without typical turbidite features; unimodal regional paleo-flow direction (versus radial pattern in fans).

Gravity flows embrace a variety of transport mechanisms in detail. They are conceptually named on the basis of the *dominant* support mechanism inferred to have been operative during final deposition (see Fig. 16). This classification involves major problems because it is *genetic*, based on inferred processes rather than objective characteristics of the deposits, and because the dominant support mechanism is uncertain in the general class

debris flow, wherein a variety of support processes are involved. The other classes named at present are *grain flow, liquefied flow*, and *turbidity current*. These four classes are ideal end-members; intermediate blends of processes are common, and different processes may dominate in different stages of transport (Fig. 17 shows interrelationships). *Rockfalls*—where particles tumble or fall from steep subaerial or subaqueous cliffs (such as reef fronts or fault scarps)—constitute a separate category of sediment gravity deposits, because they are not "flows". *Slides*, including slumps, are also not "flows" because large coherent masses are transported, but they commonly precede, accompany, or follow sediment gravity flows.

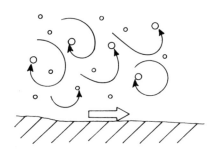

Fluid turbulence
(dispersed clasts; more rarely with concentrated clasts)

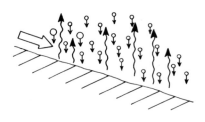

Escaping pore fluids
(high concentration but dispersed clasts)

Dispersive grain pressure
(high concentrations dispersed clasts of mixed sizes)

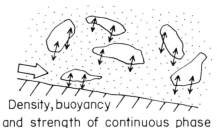

Density, buoyancy and strength of continuous phase (high concentration of clasts)

Outsize clasts supported by the combination of fluid plus all finer clasts (which may be almost wholly mud, or a full range of sizes up to largest)

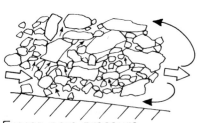

Excess pore pressure
(extreme concentration of clasts)

Interstitial fluid & fluid engulfed by advancing front of flow cannot readily escape vertically from below concentrated clasts. Engulfed water may lubricate base of flow

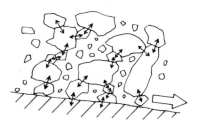

Framework stacking
(extreme concentration of clasts)

Figure 16.   Support mechanisms in sediment gravity flows.

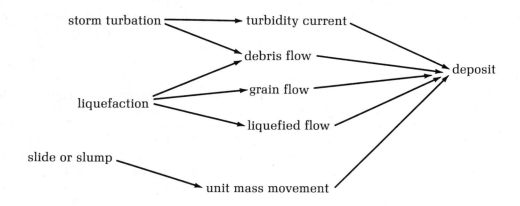

Figure 17. Common origins and interrelationships in sediment gravity flows.

TURBIDITY CURRENT

Particles in turbidity currents are supported primarily by the upward component of turbulent eddies in the flowing fluid. When deposition begins, the relatively coarse sediment is rapidly deposited as a massive poorly sorted, normally graded interval (Bouma sequence "A), which is followed rapidly by deposition of intervals "B" (parallel-laminated), "C" (cross-laminated and/or convolute-laminated) and "D" (parallel-laminated) (see Fig. 18). B, C, and D are progressively finer, and sorting thereby improves as coarser grains are selectively retained in lower intervals. Each interval overlaps deposits of the preceding interval in a down-current direc-

tion, and the "D" interval is capped by "E," which represents sediment deposited from normal "background" processes—generally fine hemipelagic muds from passive suspension. Deposition of this sequence reflects a declining flow regime, and waning traction currents may produce a very similar succession. Support mechanisms other than turbulence may also be significant in flows where the concentration of sediment is very high, and deposits may not always produce this ideal sequence (for example, "fluxoturbidites"—see Slacza and Thompson, 1981). Hence, whereas ABCD, BCD, and even CD (although with considerable faith!) vertical sequences in sedimentation units have been widely considered diagnostic of turbidities, care should be exercised in evaluating each deposit.

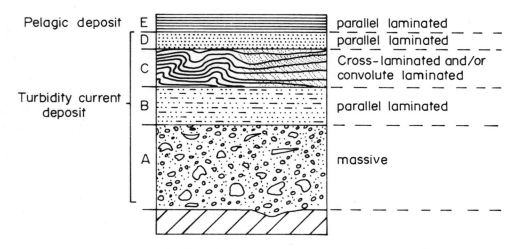

Figure 18. The Ideal Turbidite Succession and "Bouma" intervals.

GRAIN FLOWS

Grain flows consist of cohesionless particles in high concentrations moving downslope because of gravitational acceleration acting on them. They will intermittently hit each other and the upward component of the impact forces act to support the grains. This process, termed *dispersive grain pressure*, appears to operate alone as the dominant support mechanism only with sorted sand sizes on steep slopes. In the final stage of deposition, some support may be provided by upward movement of interstitial fluids as the grains settle.

LIQUEFIED FLOW

Liquified flow is the end-member case where dominant grain support is provided by inter-granular pore fluid that is escaping from beneath settling grain concentrations. For liquefaction to occur, the original fabric of a sediment must be disturbed by some trigger mechanism. Such flows are most likely initiated on relatively steep slopes; however, once moving they can flow over low gradients—although theoretically not for great distances. Liquefied flow may occur in the initial stage and in the final depositional stage of other gravity flows. Whereas at low velocities the flows are probably laminar, if they accelerate sufficiently they may flow turbulently (thereby becoming tur-bidity currents). Water expulsion structures (dish and pillar or sheet structures) are to be expected. In situations where the interstitial fluid comprises water-plus-clay (plus possibly even coarser par-ticles), the fluid will support coarser grains and/or permit longer distances of flow; in this case the support mechanism grades into that which is dis-tinctive of debris flows.

DEBRIS FLOW

Debris flow is the class of sediment gravity flow with a general mixture of grain support mecha-nisms. Laminar flow pertains during deposition and probably during much of the transportation stage. As defined by some, the dominant support mechanism during flow is the strength of the *con-tinuous phase* (water plus clays plus other rela-tively fine sediment that forms the matrix of the final deposit)—this behaves less like a true fluid and more like a low-viscosity plastic. The *dispersed phase* comprises the particles that are transported thereby—these may range from well-sorted sands to boulderly gravels with outsize clasts up to tens (even hundreds) of meters long. This definition of debris flow is simplistic because other support mechanisms, or a complex combination of them, dominate in some debris flows, and it becomes impractical to define a continuous phase if there is a complete gradation of grain sizes in the flow, all in roughly equal abundance. Dispersive grain pressure, buoyancy (enhanced by the high density and loading of the continuous phase), escaping interstitial fluids, excess pore pressure (developed where the fluid cannot readily escape upward), interclast support (in highly concentrated flows there are almost certainly instantaneous, ever-changing linkages between grains all the way to the bed of the flow)—all these may act in varying proportions in debris flows. In addition, traction flow may be involved, with the water-plus-fines acting as a high-density fluid that pushes larger clasts along the bed.

A subdivision of debris flow deposits may prove useful on the basis of objective observation of fabric. Many debris flow deposits have an abund-ance of muddy matrix, which supports randomly oriented clasts; a strong, cohesive continuous phase likely dominated during transportation. They could be termed *cohesive* or *viscous debris flows*, implying a viscous style of movement. Many other deposits have a paucity of mud and a preferred subparallel fabric in the largest clasts; during depo-sition, cohesiveness or viscosity of the flow was probably low such that intergranular processes (such as dispersive grain pressure) dominated and produced the orientation of the large clasts (either with maximum cross sections subhorizontal or poorly imbricated). Such deposits may have a sandy matrix or no true matrix; they may be clast- or matrix-supported. Some deposits comprise sorted sand deposits with diffuse parallel lamination and a sparse muddy matrix (see Lewis, 1976, 1980; Lewis et al., 1980). These could be termed *non-cohesive* or *low-viscosity debris flows*, implying a fluid of relatively low strength and viscosity.

There are some problems with this subdivision: (a) the *names* are genetic and we don't know enough about the mechanisms involved as yet; (b) a few deposits show no preferred fabric yet have very little muddy matrix (perhaps explained by a viscous-like creep action or *slurry flow* after deposition from a noncohesive flow); (c) as little as 2 wt % of clay in marine water appears to form a continuous phase with sufficient strength to transport the finest sands as debris flows (Hampton, 1975). In the system described above, such debris flows would be considered "noncohesive," yet many workers might well consider them "cohesive" because of the strength of the water-plus-clay fluid. There appears to be a predominance of the noncohesive deposits in shelf paleoenvironments (generated from shelf-type sediments) and of the cohesive deposits in submarine fan settings (generated on muddy "slope" settings). The texture of the source materials is the controlling factor, however, and deposits of either type can be found in many different settings.

Table 3 attempts to match common objective attributes of gravity flow deposits with the flow classes described above.

Table 3. Ideal End-Member Varieties of Sediment Gravity Flows

Characteristics of Deposit		Inferred Support Mechanism During Flow	Inferred Flow Type
Dominantly sand and silt in graded Bouma ABCD, BCD or CD successions. Muddy matrix and generally with mudstone interbeds. A interval poorly sorted and massive.		fluid turbulence	turbidity current
Mainly medium and finer sands and silts, generally well sorted. Dish or sheet structures, also perhaps diffuse parallel lamination. Flat bases. Deposited near to relatively steep slopes. Absence of traction-current structures.		escaping pore fluids	liquefied flow
Generally medium and finer sands and silts. Massive ungraded, or with indistinct parallel lamination or dish-structure. May contain sparse dispersed outsize clasts with sub-parallel orientation. Flat bases, though may deposit in preexisting channels. Deposited close to relatively steep slopes. Absence of traction-current structures.		dispersive grain pressure	grain flow
Generally showing extremely poor sorting with large outsize clasts, but may be well sorted fine to very fine sands with minor (to ±20%) mud matrix. Commonly deposited in channels, but may occur as constructional mounds with flat bases. Generally not graded, but may show normal, inverse, or symmetrical grading in some units. Indistinct stratification may be present— either subhorizontal or conforming to channel margins.	Mainly clast-supported or without a true matrix; may be sand matrix supported. Largest clasts with subparallel orientation. Diffuse parallel lamination in sorted sand deposits.	All of the above, plus density, buoyancy and strength of the continuous phase, plus framework stacking. Laminar flow dominates during deposition and commonly during transport. Strength of continuous phase dominant in cohesive flows, other mechanisms in noncohesive flows.	noncohesive debris flow
	Mainly matrix-supported, with matrix of mud. Clasts disorientated.		cohesive debris flow

CHEMICAL INFLUENCES

Some sedimentary minerals form in the depositional and diagenetic environments by essentially inorganic chemical processes. Detrital minerals are also variously affected by the chemistry of the weathering, transportational, depositional, and diagenetic environments. An introduction is provided here to the general factors that influence the chemistry of these environments. In later sections the evaporite and iron minerals are introduced separately as examples of chemically influenced deposits. Manganese and phosphate deposits are strongly reliant on environmental chemistry, but they are not discussed specifically in this manual. Clay mineralogy and cation adsorption properties also depend on chemical conditions, as do most diagenetic modifications to sediments (see following section). Table 4 lists a classification scheme for geochemical environments that can be applied practically from a study of sedimentary deposits.

Salinity refers to the concentration of cations and anions in solution. Per se, the concentration is slightly less than the total weight of dissolved solids per kilogram of water. It is most commonly expressed in parts per thousand (ppt), or parts per million (ppm). Salinity of normal marine waters is ±35 ppt;

hypersaline water, ±200 ppt; and brackish water, ±10 ppt. Fresh water averages about 100 ppm total salt in solution. Solubilities of the various ions differ, and the actual concentration of each in solution may vary significantly in natural waters. In sea water $Na^+ >> Mg^{2+} > Ca^{2+}$ and $Cl^- >> SO_4^{2-} > CO_3^{2-}$; many other ions are also present. Ion content and concentration in fresh water reflect the source rock in drainage catchments and climatic factors, thus vary much more than in sea water; generally $Ca^{2+} >> Na^+ > Mg^{2+}$ and $HCO_3^- >> SO_4^{2-} > Cl^-$. Salinities of interstitial waters vary depending on chemical diagenetic reactions, selective adsorption of ions on clays and colloids, and mixing as solutions migrate.

pH is the negative logarithm of the concentration of H^+ (or H_3O^+) in solution. In pure water at 20°C, H_2O dissociates to an ideal concentration of H^+ and OH^- of 10^{-7}, and this pH of 7 is taken as the neutral standard. Greater H^+ concentrations give acidic waters, pH < 7; lower concentrations give basic waters, pH > 7. (Note that pH is temperature-dependent: pure water at 120°C has a natural pH of 6. This temperature dependence becomes particularly significant in diagenesis as sediments are depressed through the geothermal gradient.) The pH of natural surface waters ranges from zero (such as in some

Table 4. A Practical Classification of Geochemical Environments by Berner (1981)

Geochemical Environment			Characteristic Authigenic Minerals
Oxic—dissolved oxygen present			Hematite, goethite, MnO_2 minerals; no organic matter.
Anoxic	Sulfidic—dissolved sulfide ($H_2S + HS^-$) present		Pyrite, marcasite, rhodochrosite, alabandite, organic matter.
	Nonsulfidic	postoxic — weakly reducing; no oxygen, sulfates not reduced	Glauconite and other $Fe^{+2} - Fe^{+3}$ silicates; siderite, vivianite, rhodochrosite. No sulfide minerals. Minor organic matter.
		methanic — strongly reducing; no oxygen, sulfates reduced with consequent methane formation	Siderite, vivianite, rhodochrosite. Earlier-formed sulfide minerals. Organic matter.

Note: This scheme is independent of salinity and pH; iron and manganese must be present in the initial sediment; microenvironments (e.g., within organism tests) may differ from the general environment; depositional and late diagenetic/epigenetic environmental conditions may produce complexly superimposed results requiring careful petrographic and paragenetic studies. The general early diagenetic sequence is: oxic ⟶ postoxic ⟶ sulfidic ⟶ methanic.

hot springs in volcanic regions) to 10 or more (some highly alkaline sea or lake waters); most natural waters have a pH between 4 and 9. Normal open marine waters have a pH of 7.5-8.5.

The importance of pH in chemical sedimentation is that it is a measure of the availability of H^+ and OH^- ions for replacing other cations and anions in minerals and for interacting with other cations/anions in solution. Dissolution, alteration, and precipitation of minerals thus strongly depend on pH. Chemical reactions involving minerals have a feedback effect on fluid chemistry and may increase or decrease pH. (For example, originally neutral water will develop a pH of 9-11 when pyriboles are abraded in it, and a pH of 5-7 with abrasion of pure quartz; hydrolyzing reactions at the feldspar/water interface may produce a local pH in excess of 11.)

Eh is a measure of the oxidation-reduction potential (volts, *v*) of a solution. *Eo* is the potential where reactions take place under standard conditions of unit activity of the reacting substances; Eh is the potential where conditions differ from this ideal (as in nature). Eo = 0.0v (neutral) is arbitrarily defined for the reaction $2H^+ + 2e \rightarrow H_2$ at 1 atmosphere pressure and in equilibrium with a solution containing H^+ at unit activity (1 molar concentration, or pH = 0)

at 25°C. Eh will vary with temperature, the concentration of reacting substances, pressure, and pH. For example, for a given reaction at 25°C, Eh will decrease by 0.06v for each unit increase of pH.

In nature, Eh generally reflects the abundance of free oxygen. Whereas a positive Eh generally gives rise to oxidizing reactions, a negative Eh does not necessarily result in reducing reactions—a reducing *capacity* is also necessary, and in sediments this is generally provided by organic matter. Where organic matter is lacking, negative Eh values may not result in reduction of materials like hematite (the iron oxide that colors sediments red).

Gaseous O_2 may be abundant (such as in the Phanerozoic atmosphere), sparse (as in many subsurface waters), or absent (for example, it may be tied up in various compounds such as CO_2, H_2O, iron and manganese oxides). Organic carbon compounds are oxidized first; if excess O_2 remains, metal cations may be oxidized. The Eh0 boundary may be above, at, or below the sediment–water interface in sedimentary systems, and O_2 may be supplied in solution late in diagenesis or during weathering.

Fig. 19 illustrates the relationship of Eh and pH to sedimentary environments and the origin of some chemical sediments.

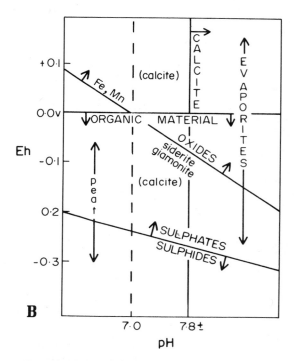

Figure 19.   Generalizations regarding Eh and pH in the sedimentary environment. A: Approximate position of some natural sedimentary environments in terms of Eh and pH. *(After Garrels and Christ,* 1965) B: Approximate fields of origin for some common sediments in terms of Eh and pH. *(After Krumbein and Garrels, 1952)*

Water presence: Water is a dipole with the two H^+ cations located $104°$ apart on the edges of the larger O^{2-} anion: The molecule has a positive and a negative side, which attract anions and cations respectively. This attraction weakens bonding within minerals, promotes hydration and hydrolysis, and facilitates dissolution.

Water circulation greatly enhances many reactions by supplying new electrolytes in solution and by removing reaction products, which otherwise tend to accumulate and retard further reaction. Circulation may occur at any stage in the history of the sediment, from a variety of causes. One of the most widespread and common causes is compaction of fine-grained sediment, which expels water laterally through adjacent lithofacies. In soils, water may move up vertically in dry periods to evaporate near the surface, where precipitation of the solutes occurs.

CO_2 content of water: CO_2 combines with H_2O to form HCO_3^- (the bicarbonate ion) and H^+ (which influences pH). It is of greatest significance in carbonate geochemistry.

Sulfide anion content: S^{2-} combines readily with metal cations (particularly iron). Most S^{2-} is produced by the breakdown of sulfate compounds, mainly through the activity of anaerobic bacteria.

Decaying *organic matter* consumes O_2, produces CO_2, liberates H^+ (thus affects pH), provides SO_4, and contains organic complexes that may be very effective in extracting metal cations from minerals and transporting them. In general, the organic content of sediments increases with decreasing grain size. In some cases, decaying organic matter may be sparse in the sediment but still be very important in microenvironments (such as within fecal pellets).

Living organisms: Passage through intestinal tracts of organisms results in abrasion and in chemical reactions; for example, both physical and chemical properties of clay-size sediments may be strongly affected. Bacterial activities may directly affect minerals (as in the breakdown of sulfate compounds) or indirectly influence reactions (as in the production of H_2S and CO_2 by anaerobic types). Aerobic bacteria attain maximum numbers at a depth of 40–60 cm in oxygenated sediment and are rapidly replaced by anaerobic types below that depth; in many claystones and lime mudstones, aerobic bacteria live in only a very thin surficial layer and in euxinic conditions they are not present at all. Bacterial content is directly related to the amount of decaying organic matter present.

Temperature: Its maximum, minimum, average and its rate of change are important in chemical reactions. Higher temperatures accelerate many chemical and biochemical reactions. Controls are climate, weather, geothermal gradient, and various internal exothermic and endothermic reactions.

DIAGENESIS

Diagenesis is the general term for all modifications to sediment characteristics that take place after the sediments have been deposited. High-temperature modifications are conceptually excluded, but the boundaries between diagenetic, hydrothermal, and metamorphic processes and effects are gradational and difficult to define strictly. The boundaries between diagenesis, pedogenesis, and weathering (including groundwater activity) are equally difficult to define; if the sediment is not subsequently transported, both initial weathering and pedogenesis can be properly considered to be part of diagenesis, whereas late-stage weathering of the sedimentary rock after it is exposed to modern surficial processes is not.

The following terms for stages of diagenesis are both simple and useful (although they are applicable essentially to marine sediments):

Syndiagenesis: refers to relatively rapid, shallow-burial modifications characterized by biological activity (such as bacterial-induced effects). The "initial" substage occurs when there is oxygen in the pore fluids, and the "early-burial" substage occurs below the zero-oxygen level ("initial" syndiagenesis will not occur at all in euxinic basins). Aerobic bacteria break down organic matter and produce carbon dioxide; anaerobic bacteria break down sulfates and produce calcite plus hydrogen sulfide, which may lead to sulfide precipitation.

Anadiagenesis: refers to longer-term (up to 10^8 yr), deeper-burial (down to 10 km) modifications such as compaction, maturation of organic compounds, most cementation, and alteration or creation of silicate minerals. Negative Eh and alkaline pH conditions prevail, but solution chemistry may vary extensively; temperature and pressure increases are major influential factors. This stage grades into metamorphism.

Epidiagenesis: refers to modifications that take place when uplift brings the sediments/sedimentary rocks into the influence of circulating fresh groundwaters. An arbitrary boundary can be set between these effects and weathering at the groundwater table level.

Diagenetic effects may be negligible or extensive, local or widespread, essentially chemical or dominantly physical. They may be obvious in hand specimen or thin section, they may be apparent only after analysis with special equipment (such as cathodo-luminescence, scanning electron microscopy, x-ray diffraction, x-ray fluorescence, electron probe), or study of a rock *sequence* may be necessary to determine them. A major indicator of the extent of diagenesis is the rank of any coaly material present in the sediments. However, rank (measured from moisture content, volatile-matter content, carbon and hydrogen content, and/or optical reflectivity) reflects essentially the degree and duration of temperature increases, and both coal diagenesis and the relationship between it and other diagenetic modifications to sediments is complex and not fully known. Fig. 20 attempts to synthesize various important diagenetic effects as a function of depth of burial.

Compositional Changes

Although diagenesis does not generally destroy the dominant detrital components of a sediment, sparse sand or finer grains of chemically unstable minerals (such as pyriboles) may be completely dissolved. Other components may be altered or replaced to such an extent that their original character is obscured—for example, feldspar or volcanic rock fragments may be so altered to clay minerals that they become difficult to distinguish from clayey sedimentary rock fragments or clayey matrix; ilmenite and other detrital titanic minerals may be altered to leucoxene; and iron-bearing minerals may be oxidized or reduced and translocated. Dissolution and alteration result from chemical interaction between the minerals and aqueous solutions that fill the interstitial pores. Solutions of differing chemistry migrate through the sediments at many stages after burial—hydrodynamic gradients result from compaction (for example, of muddy sediments deposited as a lateral facies of the unit being studied), from differential tectonism (folding or faulting), from differential heating events (such as subjacent igneous intrusion), and from other causes. In cases where units are isolated by impervious lithologies soon after burial and fluid circulation is prevented, diagenetic modifications may be insignificant (therefore, for example, early concretions display the least affected primary characteristics—although the cement of the concretion is itself a diagenetic product). In some cases, pressure or temperature changes are more significant than chemistry in causing modifications, as in the conversion of clay minerals (for example, montmorillonites $\longrightarrow$ mixed layer montmorillonite $\longrightarrow$ illite, or montmorillonite $\longrightarrow$ chlorite).

New (*authigenic*) minerals may be formed during diagenesis—not only by direct replacement of primary components, but also by direct precipitation from the interstitial fluids, which have in solution ions derived from diagenetic dissolution of other minerals (in the same or adjacent sediments) or ions inherited from the initial marine or groundwaters. Precipitated authigenic minerals tend to form euhedral crystals, but interference caused by adjacent clastic (*allogenic*) minerals or the growth of other authigenic components may prevent the development of ideal crystal forms. The authigenic components may occur as discrete, isolated new crystals or as *syntaxial overgrowths* (formed around the edges of primary "seed" crystals and commonly showing the same crystallographic/optical orientation as the parent—such as on quartz, alkali feldspars, or the single calcite crystals of echinoderm fragments) (see Fig. 21). Or they may partly to completely fill the original interstitial pore space as cements (for example, carbonate, silica, iron oxide, pyrite). Some components (such as glauconite) may form on the sea floor and be transported locally before final burial; it is convenient to use the term *perigenic* (rather than allogenic or authigenic) when such grains can be identified.

Textural Changes

Diagenetic modifications of primary rock texture may take place either as a result of chemical modifications (such as dissolution or replacement) or as a result of physical processes (such as compaction). *Pressure-solution* occurs where two grains are touching and burial (or tectonic) pressure causes one or both grains to dissolve preferentially at the points of greatest pressure, so that the grains interpenetrate and interlock (see Fig. 21). (Overgrowth interlocking commonly results in a similar texture.) When pressure solution occurs along an irregular surface marking the escape route of solutions carrying the dissolved constituents, *stylolites* form; insoluble materials tend to be concentrated along them. The net result of these processes may be to obscure original grain roundness and even original grain size. Recrystallization (mainly to larger, but occasionally to smaller, crystals) is common in carbonate sediments and may also obliterate

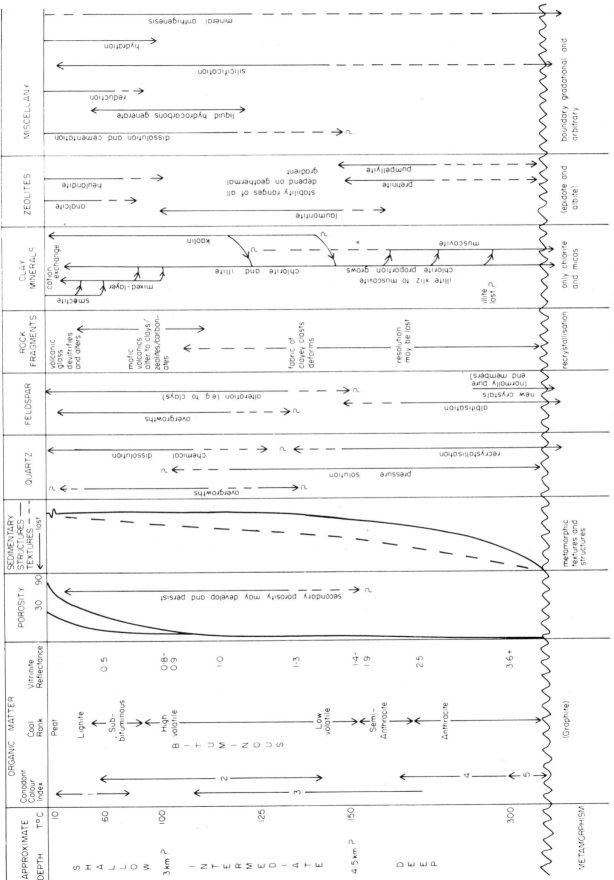

Figure 20. Generalized relationships between diagenesis and burial depth/temperature.

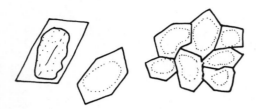

Euhedral overgrowths on rounded detrital grains; rims of quartz grains outlined by dust (e.g., iron oxides, clays).

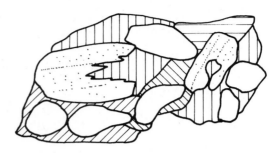

Carbonate cement (hachured) about quartz and feldspar grains (mottled). Patchy replacement of several feldspars.

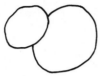

Concavo-convex contact between two grains subjected to moderate pressure-solution. At time of deposition, point or tangential contacts occur. Longitudinal contacts develop at an earlier stage.

Microstylolite between two grains subjected to extensive pressure-solution.

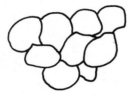

Detrital quartz grains without dust rims may develop overgrowths (and may be partly pressolved) to produce this texture. The concordant boundaries and occasional thin, irregular projections demonstrate the diagenetic modifications. Such concordance of boundaries is virtually impossible from original packing and the projections are unlikely to have survived abrasion.

Stylolite crossing many grains. Dust and insoluble minerals will ultimately concentrate along the stylolitic surface and block further movement of solutions, thereby halting the dissolution of grains on both sides.

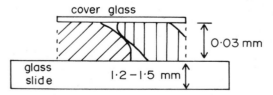

Note: Watch out for overlap relationships in thin sections, which can obscure true character of contacts.

Figure 21. Sketch examples of some diagenetic effects.

original texture. Dramatic reorganization of sediment fabric may occur from physical diagenetic processes (for example, hydroplastic to quasi-fluid deformation and dewatering structures). Burrowing activities are properly a diagenetic process as well. Changes in porosity/permeability resulting from all these modifications are particularly important to considerations of migration and entrapment of petroleum and other ore solutions.

Interpretation of diagenetic history is commonly difficult and generally requires a suite of samples from a given unit or succession, but you can and should note the overall *effects* of that history in any sample studied. The *original* characteristics of the sediment are used to infer the depositional and predepositional history, hence you must separate the diagenetic effects from the present rock character and evaluate them separately. See Larsen and Chilingar (1978) and Scholle and Schluge (1979) for excellent general discussions on diagenesis.

LITERATURE

Geological Processes

Allen, J. R. L., 1970, *Physical Processes of Sedimentation*, Allen & Unwin, London, 248p.

Bouma, A. H., D. S. Gorsline, C. Monty, and G. P. Allen (eds.), 1980, Shallow marine processes and products, *Sed. Geology* 26 (sp. issue).

Carter, R. M., 1975, A discussion and classification of subaqueous mass-transport with particular application to grain flow, slurry flow and fluxoturbidites, *Earth-Sci. Rev.* **11**:145-177.

Einsele, G., and A. Seilacher, 1982, *Cyclic and Event Stratification*, Springer-Verlag, Berlin, 536p.

Field, M., 1981, Sediment mass-transport in basins: controls and patterns, in R. G. Douglas, I. P. Colburn, and D. S. Gorsline (eds.), *Depositional Systems of Active Continental Margin Basins*, Short Course notes, SEPM Pacific Section, Society of Economic Paleontologists and Mineralogists, Tulsa, Okla., 61-84.

Hampton, M. A., 1975, Competence of fine-grained debris flows, *Jour. Sed. Petrology* **45**:834-844.

Lewis, D. W., 1976, Subaqueous debris flows of early Pleistocene age at Motunau, North Canterbury, New Zealand, *New Zealand Jour. Geology and Geophysics* **19**:535-567.

Lewis, D. W., 1980, Storm-generated graded beds and debris flow deposits with *Ophiomorpha* in a shallow offshore Oligocene sequence at Nelson, South Island, New Zealand, *New Zealand Jour. Geology and Geophysics* **23**:353-369.

Lewis, D. W., 1982, Channels across continental shelves: Corequisites of canyon-fan systems and potential petroleum conduits, *New Zealand Jour. Geology and Geophysics* **25**:209-255.

Lewis, D. W., M. G. Laird, and R. D. Powell, 1980, Debris flow deposits of early Miocene Age, Deadman Stream, Marlborough, New Zealand, *New Zealand Jour. Geology and Geophysics* **27**:83-118.

Lowe, D. R., 1979, *Sediment Gravity Flows: Their Classification and Some Problems of Its Application to Natural Flows and Deposits*, SEPM Spec. Pub. No. 27, Society of Economic Paleontologists and Mineralogists, Tulsa, Okla., 75-82.

Lowe, D. R., 1982, Sediment gravity flows: II. Depositional models with special reference to the deposits of high-density turbidity currents, *Jour. Sed. Petrology* **52**:279-297.

Middleton, G. V. (ed.), 1965, *Primary Sedimentary Structures and Their Hydrodynamic Interpretation*, SEPM Spec. Pub. No. 12, Society of Economic Paleontologists and Mineralogists, Tulsa, Okla., 265p.

Middleton, G. V., and M. A. Hampton, 1973, *Sediment Gravity Flows—Mechanics of Flow and Deposition*, Short Course Notes, SEPM Pacific Section, Society of Economic Paleontologists and Mineralogists, Tulsa, Okla., 1-38.

Middleton, G. V., and M. A. Hampton, 1976, Subaqueous sediment transport and deposition by sediment gravity flows, in D. J. Stanley and D. J. P. Swift, (eds.), *Marine Sediment Transport and Environmental Management*, Wiley, New York, 197-218.

Middleton, G. V., and J. B. Southard, 1978, *Mechanics of Sediment Movement*, SEPM Short Course No. 3. Society of Economic Paleontologists and Mineralogists, Tulsa, Okla.

Moss, A. J., 1972, Bed-load sediments, *Sedimentology* **18**:159-219.

Pierson, T. C., 1981, Dominant particle support mechanisms in debris flows at Mt. Thomas, New Zealand, and implications for flow mobility, *Sedimentology* **28**:49-60.

Slaczka, A., and S. Thompson, III, 1981, A revision of the fluxoturbidite concept based on type examples in the Polish Carpathian Flysch, *Annales Soc. Geol. Poloniae* **51**:3-44.

Stanley, D. J., and D. J. P. Swift (eds.), 1976, *Marine Sediment Transport and Environmental Management*, Wiley, New York.

Wright, L. P., 1977, Sediment transport and deposition at river mouths, a synthesis, *Geol. Soc. America Bull.* **88**:857-868.

Chemical Sedimentation

Bass Becking, L. G. M., I. R. Kaplan, and D. Moore, 1960, Limits of the natural environment in terms of pH and oxidation-reduction potentials, *Jour. Geology* **68**:243-284.

Berner, R. A., 1971, *Principles of Chemical Sedimentology*, McGraw-Hill, New York, 240p.

Berner, R. A., 1981, A new geochemical classification of sedimentary environments, *Jour. Sed. Petrology* **51**:359-365.

Cloke, P. L., 1966, The geochemical application of Eh-pH diagrams, *Jour. Geol. Education* **14**:140-148.

Degens, E. J., 1965, *Geochemistry of Sediments*, Prentice-Hall, Englewood Cliffs, N. J., 342p.

Garrels, R. M., and C. Christ, 1965, *Solutions, Minerals and Equilibria*, Harper and Row, New York, 450p.

Holland, H. D., 1976, The evolution of seawater, in B.F. Windley (ed.), *The Early History of the Earth*, Wiley, London, 559-567.

Jackson, K. S., I. R. Jonasson, and G. B. Skippen, 1978, The nature of metals-sediment-water interactions in freshwater bodies, with emphasis on the role of organic matter, *Earth-Sci. Rev.* **14**:97-146.

Krumbein, W. C., and R. M. Garrels, 1952, Origin and classification of chemical sediments in terms of pH and oxidation-reduction potentials, *Jour. Geology* **60**:1-33.

Mason, B., 1966, *Principles of Geochemistry*, 3rd ed., Wiley, New York, chap. 6, 7, 8, 9.

Rosler, H. J., and H. Lange, 1972, *Geochemical Tables*, Elsevier, Amsterdam, 468p.

Diagenesis

Aoyagi, K, and T. Kazama, 1980, Transformational changes of clay minerals, zeolites and silica minerals during diagenesis, *Sedimentology* **27**:179-188.

Brown, P. R., 1969, Compaction of fine-grained terrigenous and carbonate sediments—a review, *Canadian Petroleum Geologists Bull.* **17**:486-495.

Carey, S. W. (convenor), 1963, *Syntaphral Tectonics and Diagenesis*, symposium vol., Geology Department, University of Tasmania.

de Segonzac, G. D., 1968, The birth and development of the concept of diagenesis (1866-1966), *Earth-Sci. Rev.* **4**:153-207.

de Segonzac, G. D., 1970, The transformation of clay minerals during diagenesis and low-grade metamorphism—a review, *Sedimentology* **15**:281-346.

Friedman, G. M. 1975, The making and unmaking of limestones or the downs and ups of porosity, *Jour. Sed. Petrology* **45**:379-398.

Kossovskaya, A. G., and V. D. Shutov, 1971, Main aspects of the epigenesis problem, *Sedimentology* **15**:11-40.

Larsen, G., and G. V. Chilingar, 1978, *Diagenesis in Sediments and Sedimentary Rocks*, Developments in Sedimentology 25A, Elsevier, Amsterdam, 579p.

Runnells, D. D., 1969, Diagenesis, chemical sediments and the mixing of natural waters, *Jour. Sed. Petrology* **39**:1188-1201.

Scholle, P. A., and P. R. Schluge, (eds.), 1979, *Aspects of Diagenesis*, SEPM Spec. Pub. No. 26, Society of Economic Paleontologists and Mineralogists, Tulsa, Okla., 443p.

Siever, R., 1979, Plate-tectonic controls on diagenesis, *Jour. Geology* **87**:127-155.

Society of Economic Paleontologists and Mineralogists, 1976, *Diagenesis*, rept. series no. 1, 216p. (A selection of significant papers from *Jour. Sed. Petrology*.)

Swarbrick. E. E., 1968, Physical diagenesis: instrusive sediment and connate water, *Sed. Geology* **2**:161-175.

Teodorovich, G. I., 1961, *Authigenic Minerals in Sedimentary Rocks*, Consultants Bureau, New York, 120p.

Trurnit, P., 1968, Pressure solution phenomena in detrital rocks, *Sed. Geology* **2**:89-114.

Whetton, J. T., and J. W. Hawkins, Jr., 1970, Diagenetic origin of greywacke matrix minerals, *Sedimentology* **15**:347-361.

Sedimentary Structures

Sedimentary structures encompass a wide variety of structures formed by mechanical, chemical, and biogenic processes during deposition or shortly thereafter (Fig. 22). Most originate before sediment lithification, but some (for example, biogenic borings and some chemical concretions) are formed afterward. Because they almost invariably form where they are found, they are excellent indicators of processes operating in the depositional or early diagenetic environment. Most are best studied in the field, and it is essential to describe them comprehensively and sketch them in a notebook so that you can evaluate their implications in conjunction with other data later (and so that you do not have to return to the outcrop!). Samples with small-scale structures may warrant collection for detailed laboratory analysis; in some cases special laboratory techniques are necessary to detect or analyze them or it may be appropriate to collect acetate or latex peels from the exposure (see, for example, Bouma, 1969).

Your texts provide general introductions to the study and interpretation of sedimentary structures. Published literature is extensive; look through several of the specialist books to see the scope of these features and photographic examples (for example, Potter and Pettijohn, 1977; Pettijohn and Potter, 1964; Conybeare and Crook, 1968; Collinson and Thompson, 1982), and read some of the selected papers listed at the end of this chapter on more specific studies to see how they have been applied. Discussion and figures in this manual present only a few aspects of the field of study.

INORGANIC STRUCTURES

Inorganic sedimentary structures may be subdivided genetically into primary and secondary varieties. Primary varieties are formed at the time of deposition, either by the depositional agency (such as currents) or by tools (transported particles of any kind). Secondary varieties are formed after deposition by mechanical processes (such as brittle failure, hydroplastic deformation, or quasi-liquid flowage resulting from excess pore fluid pressures) or by chemical means (dissolution, precipitation, crystal growth, or expansion/contraction accompanying hydration/dehydration). Secondary structures generally provide information about the diagenetic environment (such as the rate of application of differential stresses, or changes in the geochemistry of pore fluids). However, in some cases they may significantly influence later sedimentation (for example, differential compaction may influence the configuration of the sediment surface; quasi-liquid instrusions may lead to extrusion at the depositional interface). Some secondary structures, such as dikes infilled from above, can have complex origins and provide information about both sedimentation and diagenesis (for example, see Lewis, 1973).

Primary sedimentary structures are important for paleoenvironmental analysis. Some indicate the direction in which sediment was moving, others indicate the orientation of sediment movement (that is, either of two directions 180° apart); both types are used to interpret paleocurrent patterns (Fig. 23) and paleogeography, as well as the nature and energy levels of depositional processes (Figs. 24 and 25, pages **42** and **43**). Some primary (and a few secondary) structures can be used to distinguish stratigraphic sequence—that is, the original younging direction (or way up); these structures must be sought in highly deformed sequences where strata may have been overturned (see Shrock, 1948). Detailed analyses of primary structures, in conjunction with textural studies, often provide

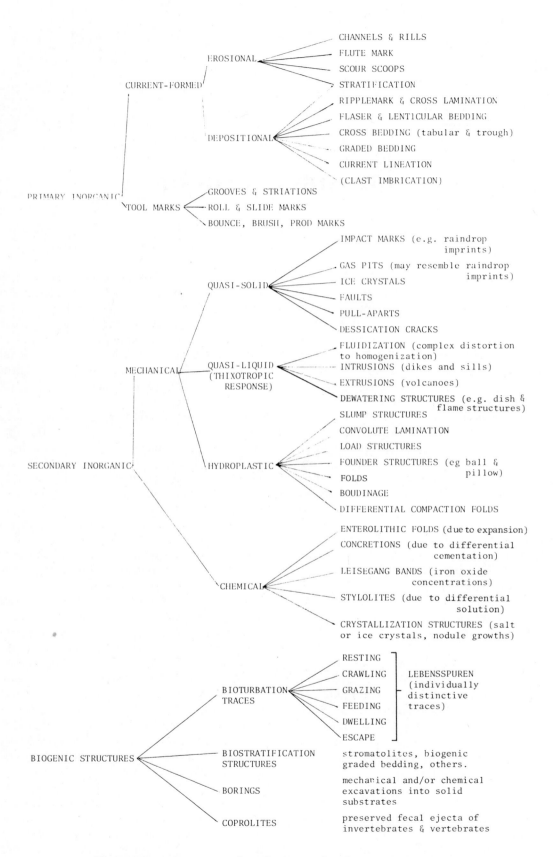

Figure 22. A genetic classification of sedimentary structures.

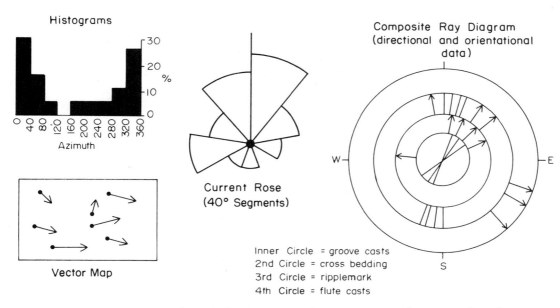

Histograms

Azimuth

Vector Map

Current Rose
(40° Segments)

Composite Ray Diagram
(directional and orientational
data)

W

E

S

Inner Circle = groove casts
2nd Circle = cross bedding
3rd Circle = ripplemark
4th Circle = flute casts

Figure 23. Examples of diagrams used to represent directional and orientational data.

other information useful in paleoenvironmental reconstruction.

Stratification, or *bedding*, is the most common primary inorganic structure; principal surfaces defining relatively large-scale, continuous units and internal stratification within these units must be considered. A standard descriptive system should be used (Fig. 26, page **44**). For rapid superficial studies, only two varieties of internal *crossbedding*—tabular and trough—are commonly recognized (see Fig. 27, page **45**), but many subvarieties can be usefully distinguished in detailed studies (Figs. 28 and 29, pages **45-47**). The direction of dip of the internal layers of crossbed sets indicates direction of flow of the paleocurrent. The angle of inclination within the sets is a function of many variables:

$$\text{angle of repose} = f\left(\frac{dm(dm - dl) \times c \times R \times d}{dl \times s \times r \times ag \times w}\right)$$

where dm = density of the sediment, dl = density of the medium of transport, d = dimension (size) of the sediment grains, R = rate of supply/unit area, w = water content of sediment, s = sphericity of grains, r = roundness of grains, c = cohesion factor, and ag = agitation factor. *Note:* These factors influence the attitude of all primary sedimentary stratification, and principal bedding planes are often not closely parallel to sea level, as is commonly assumed. The maximum angle of repose for dry angular sands is approximately 35°; postdepo-

sitional synsedimentary deformation may oversteepen or reduce the angle.

Configuration of principal stratification surfaces (bedding planes) is commonly irregular; the bedforms present range in scale from structures a few millimeters to many meters in size and vary both in shape and in mode of origin (some are purely depositional, some result from drag forces of the transporting agent on the surface, and some result from differential loading or erosion). Ripplemark is an example of a bedform with variable morphology (Fig. 30, page **48**). Hydrodynamic interpretations are possible from many of these bedforms (Fig. 24 and see Middleton, 1965; Harms et al., 1975; Middleton and Southard, 1978). In general, the original configuration represents a profile of equilibrium that reflects the balance between supply and removal of sediment. Under a given set of conditions (stable flow velocity, same rate of supply of texturally similar sediments, same width and depth of current, and so on), a single planar surface exists above which sediment continuously moves past any reference position (see Fig. 25). The sediment may be moving in continuous or discontinuous contact with the equilibrium surface as a train of transitory structures (such as ripple bedforms), but these structures will be preserved only if a rapid change of conditions permits net accumulation. Erosion eliminates many profiles of equilibrium and continuous deposition can produce some internal stratification, but almost all principal stratification surfaces represent old equilibrium profiles that

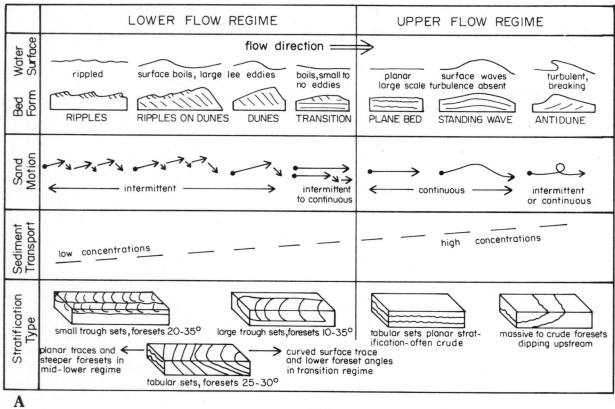

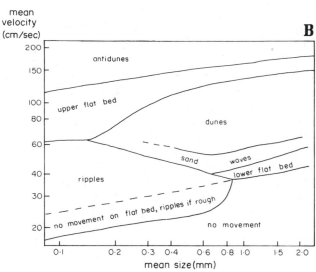

Figure 24. Form drag bedforms and internal stratification related to flow velocities and textures.
A: External and internal structures and transport of sands in relation to the flow regime. Derived for flow in alluvial channels, but widely applied as a guide to all subaqueous current systems. *(After Harms and Fahnestock in Middleton, 1965)*

B: Relationship between texture, flow velocity, and bedforms. Generalized diagram for a mean flow depth of ca. 40 cm at 10°C in flume experiments. *(After Boguchwal in Middleton and Southard, 1978)*

were preserved when conditions in the paleoenvironment changed to permit net accumulation. The temporal significance of the principal bedding planes is often obscure—most are *diastems*, reflecting relatively minor and short-term fluctuations in environmental conditions (flood events, storms, seasonal changes, normal lateral migration of current routes, and so on), but others are erosional or non-depositional disconformities of substantial temporal span.

A variety of internal primary structures are illustrated in Figure 31. Graded bedding—actually an internal fabric in sedimentation units—is illustrated in Figure 32.

Primary and secondary sedimentary structures on bedding planes are sometimes preserved as casts (American usage) or molds (British usage) on the base of more resistant lithologic units that were deposited after the structures formed. Such structures are categorized as *sole marks*.

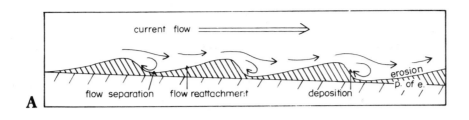

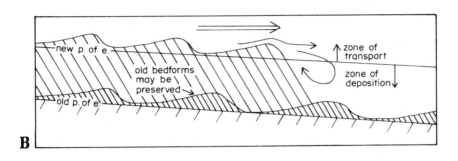

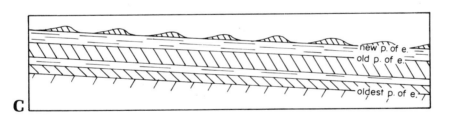

Figure 25. "Accretionary" stratification and cross-stratification. *A: Constant Flow:* bedforms migrate along a profile of equilibrium (p. of e.) with no net deposition or erosion. Because of expanding flow and flow separation over the crests of ripples or dunes, grains on the lee sides sediment under effectively zero velocity conditions and foresets are at the angle of repose. *B: Relatively large scale and/or rapid upward movement of p. of e.* (due to a rise in base level, decrease in flow velocity, increase in load or other changes): net accumulation occurs of foresets at the angle of repose, and a cross-laminated or cross-bedded unit is thereby generated between the old and new profiles of equilibrium. Continuous movement of sediment occurs above the new p. of e. *C: Slow and/or small-scale upward movement of p. of e.:* a relatively thin layer of sediment

accumulates, potentially without obvious internal structures. Each p. of e. forms a stratification plane and units may be preserved with or without obvious cross-stratification. *Downward shifts of p. of e.* (due to lowering of base level, increase of velocity, decrease of load or other change): erosion occurs above the new p. of e., which may be preserved as a stratification plane when the p. of e. subsequently moves upward.

The profiles of equilibrium concept is useful as a guide to understanding the formation of stratification in general, and in particular of major, originally subhorizontal bedding planes. Whereas erosion and deposition are rarely balanced for any lengthy period, there is a tendency for balance under any set of conditions. *(After Jopling, 1966)*

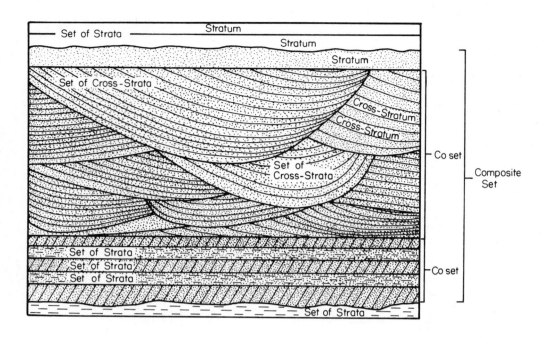

Thickness	Stratification	Cross-Stratification
120 cm	Very thick-bedded	Very thickly cross-bedded
60 cm	Thick-bedded	Thickly cross-bedded
5 cm	Thin-bedded	Thinly cross-bedded
1 cm	Very thin-bedded	Very thinly cross-bedded
2 mm	Laminated	Cross-laminated
	Thinly laminated	Thinly cross-laminated

Also note: 1. Regularity/irregularity of bedding planes. Sketch irregularities to scale.

2. Lateral persistence of beds

3. Strike and dip of bedding planes

4. Texture/composition/internal structures of the beds or strata.

Figure 26. Terminology for stratification and cross-stratification. *(After McKee and Weir, 1953)*

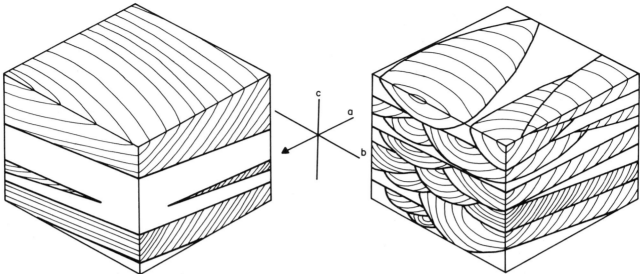

Figure 27. Tabular and trough cross-bedding. Standard reference system: *a* = direction of current flow, *c* = original vertical. *Tabular* sets have straight planar boundary surfaces, which may be either erosional or nonerosional. *Trough* sets have curved erosional boundary surfaces, and each individual cross-bed is basically trough-shaped.

The three-dimensional geometry of the sets must be known before the terminology can be applied. As with cross-lamination from assym-metrical ripples, cross-bedding can be used to infer the direction of flow (toward the azimuth of maximum dip) and the younging direction (truncated tops, tangential bases unless grain size range is restricted). To obtain directional data in trough sets, it is important to measure dips in the axial plane of the troughs. Beware of cross-sections of point bars or channel-filling deposits where apparent maximum dips may be at right angles to current flow.

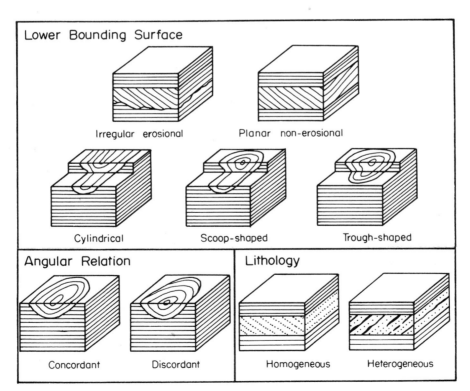

Figure 28. Detailed descriptive terminology for cross-stratification. General descriptive terms. *(After Allen, 1963) (Continues on next page.)*

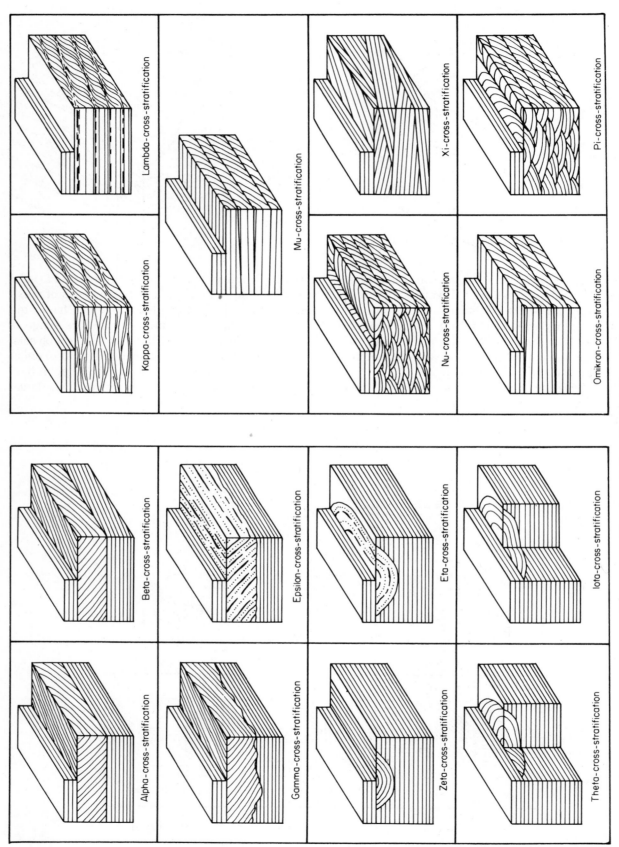

Figure 28. *(Continued)* Specific varieties. *(After Allen, 1963)*

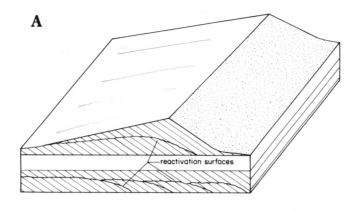

A

reactivation surfaces

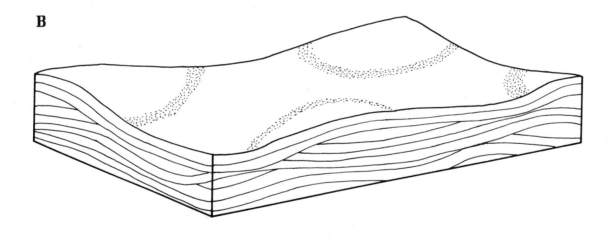

B

Figure 29.   Other "cross-bedding" structures. *A:*
Reactivation surfaces reflect interruptions in the
advance of sand waves. Erosion of the sand waves
may accompany a change in flow direction during
low river stages or tidal cycles, or a change in flow
strength. Cross-beds dip at the same angle of repose
above and below the surfaces if texture remains
constant as the flow conditions causing bedform
migration resume after each interruption.

    B: Hummocky (cross) stratification. Sets of
strata may extend for 200± m along depositional
strike, are up to 1 m thick, and may show grading.
Internal laminations are diffuse and parallel to the
hummocks and swales of the initial depositional
surface. Wavelengths of the hummocks or swales
are commonly 1–5+ m, and amplitude about 40 cm
or less; cross-cutting relationships are generally of
very low angle (<15°). *Distinctive feature:* similar
appearance in all cross-sections. *Origin:* deposition
on storm wave-influenced substrate by sediment
gravity flow initiated by nearshore storm suspension. (Amalgamated swales with few or no hummocks
characterize "swaley cross-stratification"; this
structure appears to form at slightly shallower
depths.)

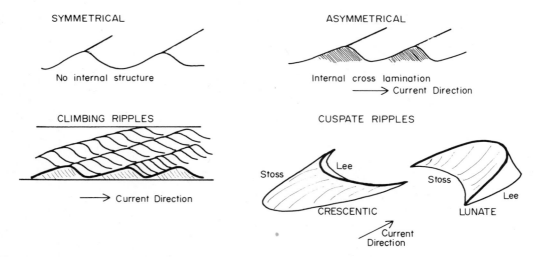

Figure 30. Some varieties of ripplemark.

Ripples result from wave or current (eolian or aqueous) activity. Current ripples theoretically are confined to bedforms produced in the lower part of the lower flow regime, but larger-scale bedforms (dunes, sandwaves, megaripples) are geometrically similar to ripples and should be described in similar ways. Ripples form only in cohesionless materials and are transitory bedforms that require either continuous fallout of sediment during formation *(ripple drift)* or a waning in the energy level of the formative process for preservation.

Symmetrical ripples generally form as a result of wave action and tend to have straight, locally bifurcating crests; wavelengths (λ, crest-to-crest or trough-to-trough lateral distances) of 0.9–200 centimeters (cm), amplitudes (A, crest-to-trough vertical distance) of 0.3–22.5 cm; and ripple indices (λ/A) of 4–13 cm, with most in the range of 6–7 cm.

Asymmetrical ripples may be formed by waves or by currents. Crests are *straight, sinuous, cuspate,* or *rhomboid*. Rhomboid ripples reflect an end member of interference patterns, where several sets of ripples form in the same area by currents flowing in different directions. Unlike crests of wave ripples, the crests of current ripples rarely if ever bifurcate, but they often terminate and are replaced by other crests. Wavelengths are usually greater than 5 cm, and are mainly 8–15 cm. The larger-

scale bedforms of similar geometry have different to overlapping quantitative values (see Reineck and Singh, 1980, and the literature list). *Flat-topped ripples* generally indicate that water depths have been reduced after ripple formation to at least the amplitude of the truncated bedform (such as in intertidal sandflats). *Herringbone cross-stratification* occurs when superposed cross-bed sets show internal beds dipping in opposite directions; it is characteristic of tidally influenced environments. *Starved ripples* are isolated ripples formed where sand or silt supplies are inadequate to form a continuous sand sheet—they appear in cross-section as lenticular bedding. *Flaser bedding* results from preservation of residual mud drapes in the troughs of ripple bedforms. (See Fig. 31 for examples of these latter varieties.)

Asymmetrical ripples indicate the direction of movement of the formative agency and the constituent sediment, whereas symmetrical ripples only indicate orientation of movement of the formative agency. Internal lamination of asymmetrical varieties can generally be used to determine younging direction (they are truncated at the top and become tangential to the principal bedding at the base if a mixture of grain sizes is available); the geometric form of symmetrical ripples can be used to determine younging direction (crests are generally sharper and narrower than troughs).

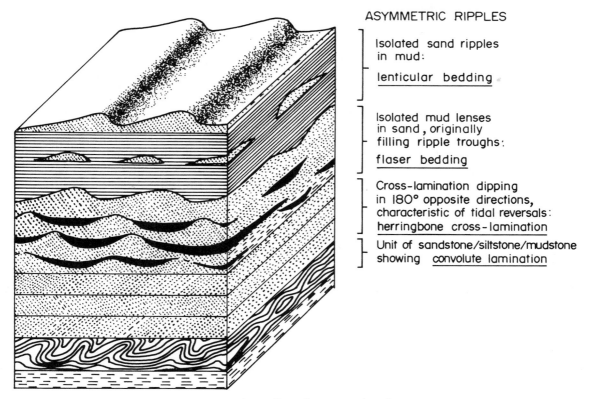

ASYMMETRIC RIPPLES

Isolated sand ripples in mud: lenticular bedding

Isolated mud lenses in sand, originally filling ripple troughs: flaser bedding

Cross-lamination dipping in 180° opposite directions, characteristic of tidal reversals: herringbone cross-lamination

Unit of sandstone/siltstone/mudstone showing convolute lamination

Figure 31. Some varieties of small-scale internal sedimentary structures.

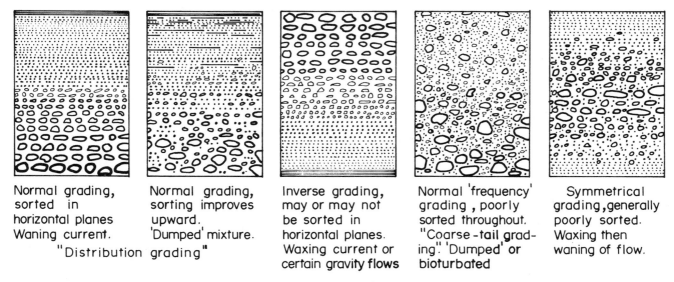

Normal grading, sorted in horizontal planes. Waning current.

"Distribution grading"

Normal grading, sorting improves upward. 'Dumped' mixture.

Inverse grading, may or may not be sorted in horizontal planes. Waxing current or certain gravity flows

Normal 'frequency' grading, poorly sorted throughout. "Coarse-tail grading". 'Dumped' or bioturbated

Symmetrical grading, generally poorly sorted. Waxing then waning of flow.

Figure 32. Some varieties of graded bedding.

BIOGENIC STRUCTURES

Biogenic structures, or trace fossils, constitute a field of study (*ichnology*) that combines the geological subdisciplines of paleontology and sedimentology. The definition given by Frey (1973)—"evidence of activity by an organism, fossil or recent, other than the production of body parts"—sounds simple, but has boundary problems in terms of paleontology. (For instance, into which field do agglutinated/precipitated shelters fall? Serpulid worm tubes are not body fossils, yet agglutinated foraminiferal tests are!) It also poses problems in sedimentology—for example, if a graded bed results from organic activity, is it then a biogenic structure? And the definition embraces a wide variety of fundamentally different structures, such as *bioturbation structures* (disruption of physical structures or sediment fabrics), *biostratification structures*

(such as stromatolites), *bioerosion structures* (such as borings), and even *coprolites* (see Hantzschel et al., 1969). The literature is extensive, particularly since the mid-sixties; among those in the selected list given here, start by reading Frey (1973) or the first chapter in Frey (1975).

Burrow structures are most widely distributed and tend to be the most widely used for sedimentological interpretations. They are *in situ* results of animal behavior patterns. Although the creator organism cannot be identified in most cases, deduction of the stimuli that led to a particular behavior can provide useful information about the paleoenvironment and conditions of sedimentation. Particular assemblages of traces (mostly burrows) characterize similar environments throughout the Phanerozoic (Fig. 33). (Akin to the concept of homeomorphy in paleontology, similar traces are created by different organisms if the stimuli are similar.)

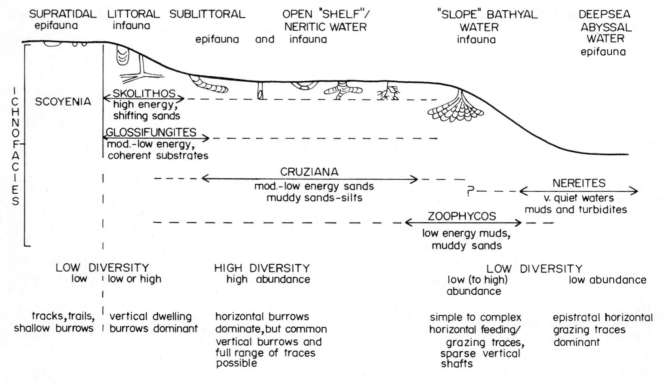

Figure 33. Environmental zonation of trace fossils. Physical, chemical, and biological factors establish ecological habitats that may be similar at different water depths or distances from shorelines. A generalized ichnofacies model is presented for an idealized profile; the namegivers may not be present,

but analogous forms must dominate for the "ichnofacies" to be named. The *assemblage* of traces determines the ichnofacies; some *lebensspuren* (even the namegivers) may occur outside the facies of which they are characteristic.

Burrows are classified in three ways. Their physical characteristics must be described (see Fig. 34, and table 3.2 of Simpson in Frey, 1975). The kind of behavior that produced the structure is commonly inferred (see Fig.35). And for the individually distinctive traces, or *Lebensspuren, form* genera and *form* species are defined on the basis of the trace morphology, following as closely as feasible the system for body fossils of the International Commission on Zoological Nomenclature. For example, *Ophiomorpha nodosa* identifies a characteristic knobbly, cylindrical burrow form. The name of the creator organism is not used, because the creator is often not known, because the same organism can create distinctively different traces, and because different organisms can create similar traces. Hence an *Ophiomorpha* trace may connect to a trace named *Thalassinoides*, indicating that the same organism created the distinctively different trace forms. Or *Thalassinoides* may be present in settings where *Ophiomorpha* creators appear to have been absent. In some cases, substantial variation in trace morphology must be permitted within the same form genus or species because of the creator's behavioral eccentricities; also, the same trace species, defined on the same basis, can belong to several different genera. There are additional ways in which the formal classification system differs from that for body fossils. Part W of the *Treatise on Invertebrate Paleontology* (Hantzschel, 1975) provides the most comprehensive index of *Lebensspuren*, but new forms have been defined in the literature since 1975.

When working with traces that are not immediately recognized, it is essential that you establish their full three-dimensional configuration, including any morphological variation that appears to reflect minor behavioral eccentricities. Then carefully check the literature, beginning with Hantzschel (1975) before you assign a new formal name. Note carefully all relationships to other physical sedimentary structures and any body fossils that may occur in the same sequence.

When interpreting the history of burrow assemblages, note that the youngest trace (the one that cuts the others) generally reflects burrowing by the organism that operated at the greatest depth within the sediment. As with epifauna, infaunal communities are trophic: different organisms with different needs operate at different depths. Hence the effects of each deeper operator will be progressively superimposed on the effects of the shallower forms as sedimentation continues. Exceptions occur

when penecontemporaneous erosion lowers the sediment–water interface and shallower forms operate within originally deeper layers.

An understanding of the fundamental needs of organisms, as well as of physiochemical characteristics of sedimentary environments, is necessary for ethological interpretation of trace fossils and general evaluation of the paleoenvironmental setting in which the traces formed. For example, the mere presence of benthic traces indicates oxygenated bottom waters, since organisms must respire. In general, vertical burrows dominate where high-energy conditions keep the surficial sediment mobile (such as on beaches and in bar-front settings), whereas horizontal burrows dominate where energy conditions are low and nutrients are abundant within the substrate. Suspension feeders dominate where substrates are firm and there is little suspended sediment in the water, whereas sediment-eaters dominate where the substrate is soft and includes abundant nutrients. Below the sediment–water interface, at various but generally shallow depths, conditions become anaerobic or anoxic. To respire, animals operating in such an evironment must maintain a connecting shaft to the overlying oxygenated waters. Many organisms, even under aerobic substrate conditions, maintain a continuously open tube (basically U-shaped) with two connections to the water to aid circulation and to eject fecal matter. In traces such as *Diplocraterion* (Fig. 34), the U-tube is characteristic of the trace, but in traces such as *Ophiomorpha* (a distinctive arthropod burrow), the basic U-tube geometry cannot be determined in typical exposures and became apparent only from studies of the modern burrow systems. Only one vertical shaft can be seen in most *Zoophycos* traces, but a complex open U-tube must have been maintained by the deep-burrowing creator, as sketched in Figure 34.

Trace fossil studies have been applied to a variety of geological interpretations other than those mentioned above, for example:

They indicate the presence of otherwise unpreservable soft-bodied organisms and, in rare cases, the soft-part morphology of extinct organisms or the way in which appendages operated.

Some have demonstrated the different life habits of extinct organisms, for example the lack of small trilobite tracks and trails has convincingly shown the planktonic/nektonic habit of juveniles of some genera.

Figure 34. Some descriptive terminology for trace fossils.

Toponomic Classification* (mode of preservation and occurrence)	
Character	Name
Trace occurs on top surface of stratum	epichnia or epirelief structure—concave groove or convex ridge
Trace occurs on bottom of stratum	hypichnia or hyporelief—convex ridge or concave groove
Trace is within stratum and infilled with similar sediment	endichnia ⎫ full
Trace is within stratum and filled with different sediment to that surrounding it	exichnia ⎭ relief structures

*For other descriptive schema, see Hantzschel (1975) and Simpson (in Frey, 1975, Table 3.5)

Meniscus structure: sets of curved lamellae or small ridges between principal boundaries of the trace. Some represent packing of sediment behind organisms that moved in the sediment *(active backfill),* others represent *passive fill* of abandoned burrows. Most are consistently concave in the direction that the creator organism was moving.

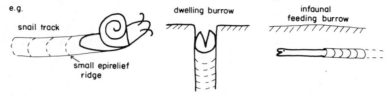

Spreiten are the composite traces resulting from multiple activities of the creator organism— e.g., numerous feeding forays—each overlapping or connected to a central burrow. The final result may not resemble any single episode of activity.

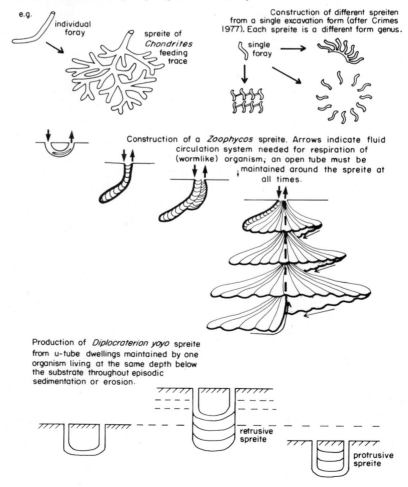

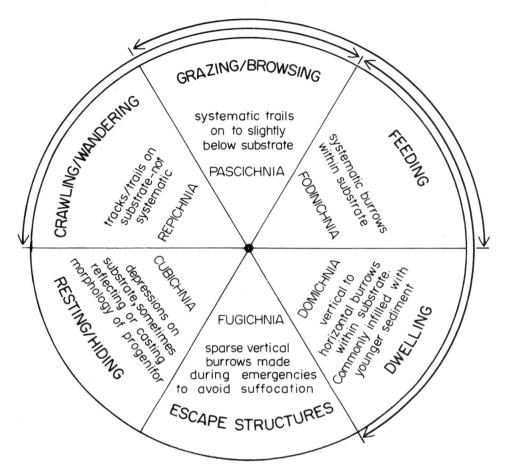

Figure 35. Ethological (functional) classification of trace fossils (based on the system of A. Seilacher; see Frey, 1973, or Simpson, chap. 3 in Frey, 1975). External arrows show possible overlap of functional classes.

Note: Application of this terminology requires knowledge or inference of the behavior of the creator organism, and is complicated by the fact that organisms may produce very similar traces when performing different functions, or use the same burrow for different functions (e.g., polychaetes feeding from a dwelling). See Schafer (1972) for guidelines to the behavior of organisms.

Forms such as *Diplocraterion yoyo* (or merely the different relative intensities of bioturbation in a sequence) indicate rates and nature of sedimentation. The presence in burrows of sediment that is absent in the overlying strata often indicates episodic erosion in the paleoenvironment.

Where the trace shows details of the organisms' morphology (that is, it mirrors biologic speciation), it may be useful in stratigraphic correlation.

They may be relevant to the interpretation of diagenesis, because burrowing disrupts the original sediment fabric, mucoid burrow linings may localize later chemical precipitation, and because the activity of the trace-producers influence the local chemistry of the sediment and pore water.

They may indicate aspects of the physical character of the sediment, for example whether or not it was sufficiently cohesive to maintain open tubes. In conjunction with the pattern and type of burrows present, this information may suggest the presence of discontinuity surfaces (breaks of greater duration than diastems but lesser than unconformities) or disconformities (Fig. 36).

Actualistic studies indicate that burrowing organisms inhabit most marine environments. Absence of fossil burrows in many stratigraphic intervals is thus puzzling, since substantial changes in the marine environment during Phanerozoic time are not likely. Absence in marine sediments may result from euxinic or anoxic bottom waters; hypersaline waters; a very mobile substrate (and perhaps some sloppy, soupy substrates); too much suspended

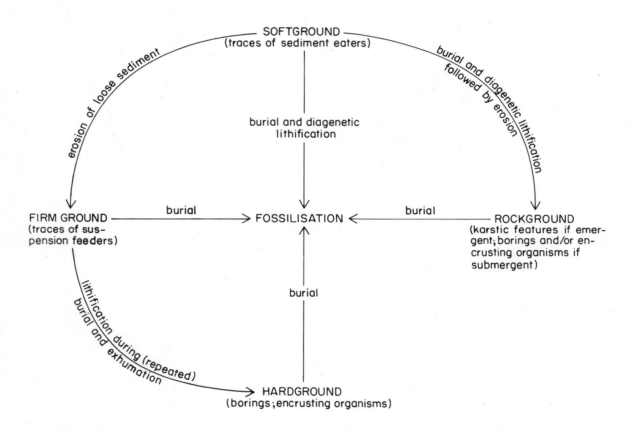

Figure 36.   A classification of substrates (after A. Seilacher, pers. comm., 1982).

Softground burrow fillings differ from surrounding sediment only slightly, due to effects of ingestion, concentration of organic matter in fecal packets (e.g., pellets), and possibly from differential diagenesis (resulting from physiochemical differences caused by biological activity).

Firmground dwelling burrows and *hardground or rockground* borings are filled by sediment that was deposited *on* the substrate surface; erosion may have removed that sedi-

ment from the surface prior to final burial. The fill may be similar to, or very different from, that surrounding the traces. The age difference between the filling and surrounding sediment may not be detectable by independent evidence such as paleontology or absolute-age determination (even rockgrounds can form rapidly under certain combinations of texture, composition, and diagenetic conditions).

sediment (unlikely to persist throughout prolonged periods); or too rapid sedimentation (although a high average sedimentation rate of 10 cm/1000 yr is not too rapid for bioturbation). Other evidence in the sediments commonly exists to indicate each of these settings, yet many barren strata do not have such evidence. In such cases, it is tempting to rely on an explanation of episodic sedimentation, whereby exceptionally powerful events cause rapid redeposition of sediments in layers too thick to be penetrated fully by any subsequently reestablished infaunal community. Repetition of the rare, high-magnitude events during the gradual aggradation of sediment may result in preferential preservation of barren strata that are not representative of the average environmental conditions.

LITERATURE

Inorganic Sedimentary Structures

GENERAL

Allen, J. R. L., 1967, Notes on some fundamentals of paleocurrent analysis with reference to preservation potential and sources of variance, *Sedimentology* 9:75-88.

Allen, J. R. L., 1982, *Sedimentary Structures. Their Character and Physical Basis*, Developments in Sedimentology No. 30A, Elsevier, Amsterdam, 593p.

Bouma, A. H., 1969, *Methods for the Study of Sedimentary Structures*, Wiley, New York, 458p.

Conybeare, C. E. B., and K. A. W. Crook, 1968, *Manual of Sedimentary Structures, Bur. Mineral Resources, Geol. & Geophys. Bull.* 102, 327p.

Dott, R. H. Jr., 1974, Paleocurrent analysis of severely deformed, flyschtype strata—a case study from South Georgia Island, *Jour. Sed. Petrology* 44:1166-1173.

Dzulynski, S., and E. K. Walton, 1965, *Sedimentary Features of Flysch and Greywackes*, Developments in Sedimentology No. 7, Elsevier, Amsterdam, 274p.

Elliott, R. E., 1965, A classification of subaqueous sedimentary structures based on rheological and kinematical parameters, *Sedimentology* 5:193-209.

Harms, J. C., J. B. Southard, D. R. Spearing, and R. G. Walker, 1975, *Depositional Environments as Interpreted from Primary Sedimentary Structures and Stratification Sequences*, SEPM Short Course No. 2, lecture notes, Society of Economic Paleontologists and Mineralogists, Tulsa, Okla.

Klein, G. de V., 1967, Paleocurrent analysis in relation to modern marine sediment dispersal patterns, *Am. Assoc. Petroleum Geologists Bull.* 51:366-382.

Middleton, G. V. (ed.), 1965, *Primary Sedimentary Structures and Their Hydrodynamic Interpretation*, SEPM Spec. Pub. No. 12, Society of Economic Paleontologists and Mineralogists, Tulsa, Okla., 265p. (Also read paper by Harms and Fahnestock, pp. 84-115.)

Middleton, G. V., and J. B. Southard, 1977, *Mechanics of Sediment Movement*, SEPM Short Course No. 3, lecture notes (esp. chap. 7), Society of Economic Paleontologists and Mineralogists, Tulsa, Okla.

Pettijohn, F. J., and P. E. Potter, 1964, *Atlas and Glossary of Primary Sedimentary Structures*, Springer-Verlag, New York, 370p.

Potter, P. E., and F. J. Pettijohn, 1977, *Paleocurrents and Basin Analysis*, 2nd ed., Springer-Verlag, New York, 425p.

Ramsay, J. G., 1961, The effects of folding upon the orientation of sedimentation structures, *Jour. Geology* 69:84-100.

Reineck, H. -E., and I. B. Singh, 1980, *Depositional Sedimentary Environments*, 2nd ed., Springer-Verlag, New York, 549p.

Shrock, R. R., 1948, *Sequence in Layered Rocks*, McGraw-Hill, New York, 507p.

BEDDING/CROSS-BEDDING

Allen, J. R. L., 1963, The classification of cross-stratified units, with notes on their origin, *Sedimentology* 2:93-114. (Also see discussion by K. A. W. Crook and reply in *Sedimentology* 3:249-254.)

Hamblin, W. K., 1962, X-ray radiography in the study of structures in homogeneous sediments, *Jour. Sed. Petrology* 32:201-210.

Jopling, A. V., 1966, Some applications of theory and experiment to the study of bedding genesis, *Sedimentology* 7:71-102.

Kocurek, G., and R. H. Dott, Jr., 1981, Distinctions and uses of stratification types in the interpretation of eolian sand, *Jour. Sed. Petrology* 51:579-595.

Kuenen, P. H., 1953, Significant features of graded bedding, *Am. Assoc. Petroleum Geologists Bull.* 37:1044-1066.

McBride, E. F., R. G. Shepherd, and R. A. Crawley, 1975, Origin of parallel, near-horizontal laminae by migration of bed forms in a small flume, *Jour. Sed. Petrology* 45:132-139.

McKee, E. D., 1967, Tabular and trough cross bedding: comparison of dip azimuth variability, *Jour. Sed. Petrology* 37:80-86.

McKee, E. D., and G. W. Weir, 1953, Terminology for stratification and cross-stratification in sedimentary rocks, *Geol. Soc. America Bull.* 64:381-390.

RIPPLEMARK

Allen, J. R. L., 1968, *Current Ripples*, North Holland, Amsterdam, 433p.

Clifton, H. E., 1976, Wave-formed sedimentary structures—a conceptual model, in R. A. Davis and R. L. Ethington (eds.), *Beach and Nearshore Sedimentation*, SEPM Spec. Pub. No. 24, Society of Economic Paleontologists and Mineralogists, Tulsa, Okla., pp. 126-148.

Harms, J. C., 1969, Hydraulic significance of some sand ripples, *Geol. Soc. America Bull.* **80**:363-396.

Jopling, A. V., and R. G. Walker, 1968, Morphology and origin of ripple-drift cross-lamination, with examples from the Pleistocene of Massachusetts, *Jour. Sed. Petrology* **38**:971-984.

Reineck, H.-E., and F. Wunderlich, 1968, Classification and origin of flaser and lenticular bedding, *Sedimentology* **11**:99-104.

Tanner, W. F., 1967, Ripple mark indices and their uses, *Sedimentology* **9**:89-104.

MISCELLANY

Boswell, P. G. H., 1948, The thixotropy of certain sedimentary rocks, *Sci. Progress* **36**:412-422. (Also see Boswell, 1951, The trend of research on the rheotropy of geological materials, *Sci. Progress* **39**:608-622.)

Dionne, J.-C., 1973, Monroes: a type of so-called mud volcanoes in tidal flats, *Jour. Sed. Petrology* **43**:848-856.

Freeman, P. S., 1968, Exposed middle Tertiary mud diapirs and related features in south Texas, *Am. Assoc. Petroleum Geologists Mem.* **8**:162-182.

Gregory, M. R., 1969, Sedimentary features and penecontemporaneous slumping in the Waitemata Group, Whangaparaoa Peninsula, North Auckland, New Zealand, *New Zealand Jour. Geology and Geophysics* **12**(1):248-282.

Lewis, D. W., 1973, Polyphase limestone dikes in the Oamaru region, New Zealand, *Jour. Sed. Petrology* **43**:1031-1045.

Lowe, D. R., and R. D. LoPiccolo, 1974, The characteristics and origins of dish and pillar structures, *Jour. Sed. Petrology* **44**:484-501.

McKee, E. D., and M. Goldberg, 1969, Experiments on formation of contorted structures in mud, *Geol. Soc. America Bull.* **80**:231-244.

Moore, D. G., 1961, Submarine slumping, *Jour. Sed. Petrology* **31**:344-357.

Newsom, J. F., 1903, Clastic dikes, *Geol. Soc. America Bull.* **14**:227-268.

Plummer, P. S., and V. A. Gostin, 1981, Shrinkage cracks: dessication or synaeresis, *Jour. Sed. Petrology* **51**:1147-1156.

Ramberg, H., 1955, Natural and experimental boudinage and pinch and swell structures, *Jour. Geology* **63**:512-526.

Rattigan, J. H., 1967, Depositional, soft sediment, and post-consolidation structures in a Palaeozoic aqueoglacial sequence, *Geol. Soc. Australia Jour.* **14**:5-18.

Sanders, J. E., 1960, Origin of convolute lamination, *Geol. Mag.* **97**:409-421.

Woodcock, N. H., 1979, The use of slump structures as palaeoslope orientation estimators, *Sedimentology* **26**:83-99.

Biogenic Structures

Basan, P. B. (ed.), 1978, *Trace Fossil Concepts*, SEPM Short Course No. 5, Society of Economic Paleontologists and Mineralogists, Tulsa, Okla., 201p.

Budd, D. A., and R. D. Perkins, 1980, Bathymetric zonation and paleoecological significance of microborings in Puerto Rican Shelf and slope sediments, *Jour. Sed. Petrology* **50**:881-904.

Carriker, M. R., L. H. Smith and E. T. Wilce, (eds.), 1969, Penetration of calcium carbonate substrates by lower plants and invertebrates, *Am. Zoologist* **9**:629-1020.

Chamberlain, C. K., 1971, Bathymetry and paleoecology of Ouachita geosyncline of Southeastern Oklahoma as determined from trace fossils, *Am. Assoc. Petroleum Geologists Bull.*, **55**:34-50.

Crimes, T. P., 1968, *Cruziana*: a stratigraphically useful trace fossil, *Geol. Mag.* **105**:360-364.

Çrimes, T. P., 1977, Modular construction of deep water trace fossils from the Cretaceous of Spain, *Jour. Paleontology* **51**:591-605.

Crimes, T. P., and J. C. Harper (eds.), 1971, Trace fossils, *Geol. Jour., Spec. Issue No. 3*, 547p.

Crimes, T. P., and J. C. Harper (eds.), 1977, Trace fossils 2, *Geol. Jour., Spec. Issue No. 9*, 351p.

Dapples, E. C., 1938, The sedimentational effects of the work of marine scavengers, *Am. Jour. Sci.* **36**:54-65.

Ekdale, A. A. (ed.), 1978, Trace fossils and their importance in paleoenvironmental analysis, *Palaeogeography, Palaeoclimatology, Palaeoecology* **23** (spec. issue):167-373.

Frey, R. W., 1973, Concepts in the study of biogenic sedimentary structures, *Jour. Sed. Petrology* **43**:6-19.

Frey, R. W. (ed.), 1975, *The Study of Trace Fossils*, Springer-Verlag, New York, 562p.

Frey, R. W., J. D. Howard, and W. A. Pryor, 1978, *Ophiomorpha*: its morphologic, taxonomic, and environmental significance, *Palaeogeography, Palaeoclimatology, Palaeoecology* **19**:199-229.

Hantzschel, W., 1975, *Trace Fossils and Problematica*, 2nd ed. Treatise on Invertebrate Paleontology, Part W, Miscellanea, Supplement 1, C. Teicher, ed., Geological Society of America and University of Kansas, Boulder, Colo., and Lawrence, Kan., 269p.

Hantzschel, W., F. el-Baz, and G. C. Amstutz, 1969, *Coprolites, an Annotated Bibliography*, Geol. Soc. America, Mem. 108, 132p.

Howard, J. D., 1968, X-ray radiography for examination of burrowing in sediments by marine invertebrate organisms, *Sedimentology* **11**:249-258.

Howard, J. D., 1972, Trace fossils as criteria for recognizing shorelines in stratigraphic record, in J. K. Rigby and W. K. Hambin (eds.), *Recognition of Ancient Sedimentary Environments*, SEPM Spec. Pub. No. 16, Society of Economic Paleontologists and Mineralogists, Tulsa, Okla., 215-225.

Howard, J. D., H.-E. Reineck, and S. Rietschel, 1974, Biogenic sedimentary structures formed by heart urchins, *Senckenbergiana Maritima* **6**:185-201.

Kern, J. P., and J. E. Warme, 1974, Trace fossils and bathymetry of the Upper Cretaceous Point Loma Formation, San Diego, California, *Geol. Soc. America Bull.* **85**:893-900.

Lessertisseur, J., 1955, Traces fossiles d'activité animale et leur signification paleobiologique, *Soc. Géol. France Mém. 75* (ns), 150p.

Lewis, D. W., 1970, The New Zealand *Zoophycos, New Zealand Jour. Geology and Geophysics* **13**:295-315.

Redfield, A. C., 1958, The biological control of chemical factors in the environment, *Am. Scientist* **46**:205-221.

Risk, M. J., and J. S. Moffat, 1977, Sedimentological significance of fecal pellets of *Macoma bathica* in the Minas Basin, Bay of Fundy, *Jour. Sed. Petrology* **47**:1425-1436.

Schafer, W., 1972, *Ecology and Palaoecology of Marine Environments,* Oliver & Boyd, Edinburgh, 568p.

Seilacher, A. 1964, Biogenic sedimentary structures, in J. Imbrie and N. Newall (eds.), *Approaches to Paleoecology,* Wiley, New York, 296-316.

Seilacher, A. 1967, Fossil behaviour, *Sci. American* **217**:72-80.

Simpson, S., 1957, On the trace-fossil Chondrites, *Geol. Soc. London Quart. Jour.* **112**:475-499.

Tevesz, M. J. S., F. M. Soster, and P. L. McCall, 1980, The effects of size-selective feeding by oligochaetes on the physical properties of river sediment, *Jour. Sed. Petrology* **50**:561-568.

Winder, C. G., 1968, Carbonate diagenesis by burrowing organisms, *23rd Internat. Geol. Congress* **8**:173-183.

Texture of Detrital Sediments

The term *texture* refers to the size, shape (sphericity or form), roundness, grain surface features, and fabric (packing and orientation) of grains. (See Pettijohn, 1975, Ch. 3, for a good general discussion.) Of these properties, size distributions of detrital sediments have been most widely and intensively studied to provide data for interpreting geological history, hence this chapter emphasizes a size classification system, quantitative (statistical) size analysis, and methods for obtaining size data.

CLASSIFICATION SYSTEMS

Systematic description of rocks is necessary both for the orderly study of rock suites and the communication of observations to others. Many systems are in use and various criteria or class limits are frequently applied to the same term. Hence, in any paper dealing in detail with rocks, the system of terminology used must be specified, else confusion reigns! The system presented here for detrital sediment texture (and composition) is that of Folk, Andrews, and Lewis (1970)—see that publication for an expanded presentation.

It is important that organization of the textural classification be related to, and expressed in terms of, a generally accepted grain size scale. The scale most commonly used for sediments by English-speaking geologists is Udden's 1898 grade scale as slightly modified by Wentworth (1922). This scale is geometric (each boundary is ½ or 2 × the millimeter value of the next). To avoid dealing with fractions of millimeters and having to use $\log_2$ graph paper, W. C. Krumbein (1934) introduced the more convenient $\varnothing$ (phi) scale, where $\varnothing = -\log_2$ mm. These scales, and their verbal terminology, are

given in Fig. 37. (See Tanner, 1969, for a discussion of the advantage of this grade scale in sediment studies and for comparison with other size scales).

It is widespread practice to subdivide the range of detrital grain sizes into three groups: gravel, sand, and mud. Gravels mostly derive from blocky fracturing of the source rocks; sands result from the breakdown of rocks into their component crystals (numerous exceptions); and muds represent the finest products of disintegration (silt and clay-size detritus) plus the products of decomposition (the clay minerals). Silt and clay also generally behave differently in sedimentation. Thus it is also reasonable to make a major distinction between silt and clay sizes in the mud fraction.

The grain size distribution of any detrital sediment can be represented on one of two triangular diagrams. Use of triangles permits the graphical plotting of data for sample comparisons. The relative proportions of gravel, sand, and mud are plotted on one triangle; sediment without gravel is plotted on the other. Categories are named according to the relative proportions of sand, silt, and clay. By providing additional information, such as the "average" size of gravel and sand, samples within one textural category can be distinguished.

Gravel-Bearing Sediments

Gravel-bearing sediments consist of gravel (grains larger than 2 mm in diameter), with or without sand (grains 2.0–0.0625 mm in diameter) and/or mud (grains smaller than 0.0625 mm in diameter). The gravel:sand:mud ratio in a sample is determined to classify the sediment, and percentages are plotted

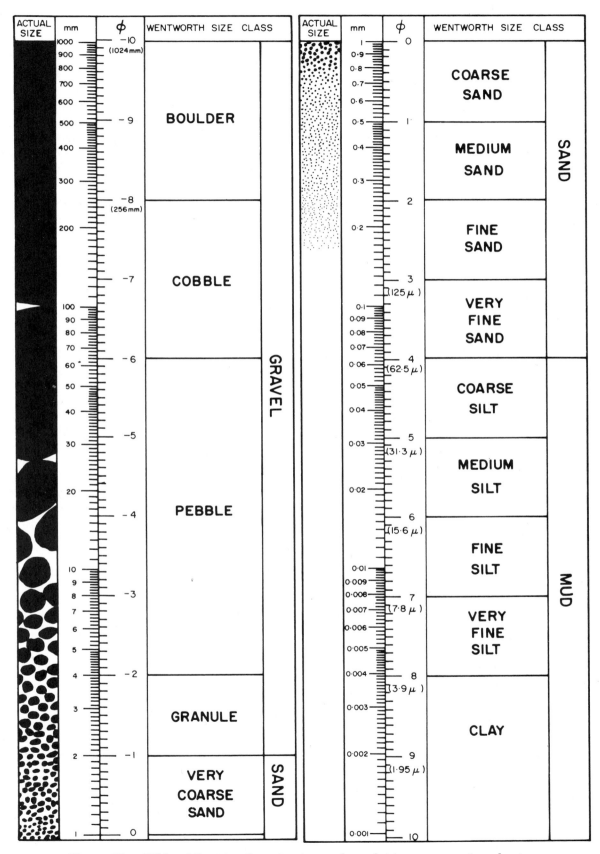

Figure 37. Udden-Wentworth grain-size scale and ϕ/mm conversion chart.

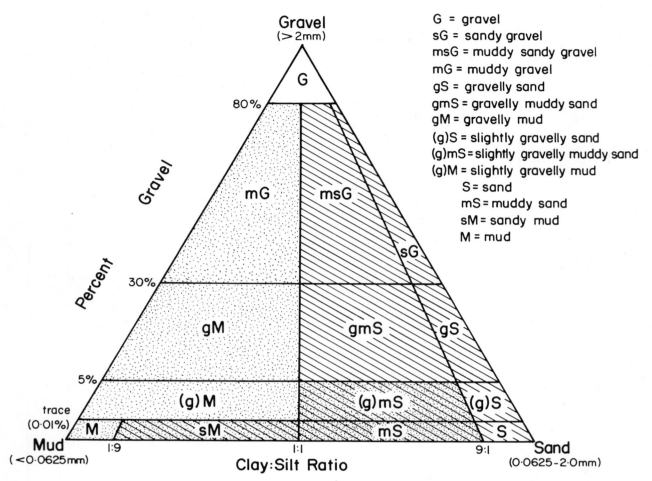

Figure 38. Textural terminology for gravel-bearing detrital sediments. Specify modal size of gravel throughout. Specify modal size of sand in cross-hatched area. Where practicable, substitute "silty" or "clayey" for "muddy" in stippled area. Specify "clast-supported" or "matrix-supported" when sand or mud fraction is abundant. *(After Folk, Andrews, and Lewis, 1970)*

on a triangular diagram (Fig. 38), the apices of which represent 100% gravel, 100% sand, and 100% mud, respectively. The triangle is subdivided into 14 textural classes.

It is a simple procedure to determine the textural class to which a sample belongs. First determine the percentage of gravel. Five categories are represented by tiers in the triangle: more than 80%, 30–80%, 5–30%, trace (0.01%)–5% gravel, and gravel-free. The last category is provided so that the small number of gravel-free samples in a suite of gravelly sediments may be plotted on the same diagram. Next determine the ratio of sand to mud. Boundaries

at 9:1 and 1:1 sand to mud subdivide each of the middle tiers of the triangle into three named classes; the upper tier consists of one class only, the gravel class. The bottom (gravel-free) tier is subdivided at 9:1, 1:1, and 1:9 sand to mud to give four textural classes.

In any type of study, the textural class should be determined in the field. It is difficult to collect representative samples of gravelly sediment for laboratory analysis. Where sand or mud is abundant in gravelly sediments, it is useful to note whether the deposit has a *clast-supported* or *matrix-supported* texture.

Gravel-Free Sediments

Size distributions of gravel-free sediments and sedimentary rocks are also plotted on a triangular diagram (Fig. 39) according to the proportions of sand (2.0–0.0625 mm), silt (0.0625–0.0039 mm), and clay (less than 0.0039 mm). The apices of the triangle are then 100% sand, 100% silt, and 100% clay, respectively. The triangle is subdivided to give 10 textural classes.

Again the procedure for determining the textural class is simple. First determine the percentage of sand in the sample. Four categories are represented by tiers on the triangle: more than 90%, 50–90%, 10–50%, and less than 10% sand. Next determine the ratio of silt to clay. Boundaries at 2:1 and 1:2 silt to clay subdivide each of the three lower tiers of the triangle into three classes. The upper tier consists of one class only, the sand class.

Usage

Only loose sediment names have been given in the illustrations. When the sediment is indurated, substitute *conglomerate* (or *breccia*, if particles are angular) for *gravel*, and substitute *sandstone*, *mudstone*, *siltstone*, and *claystone* for *sand*, *mud*, *silt*, and *clay*.

The preceding discussion should show that you need only determine the percentage *range* of a size fraction and not the precise percentage to assign a sample to its textural class. Thus the nomenclature may be applied in the field by using nothing more

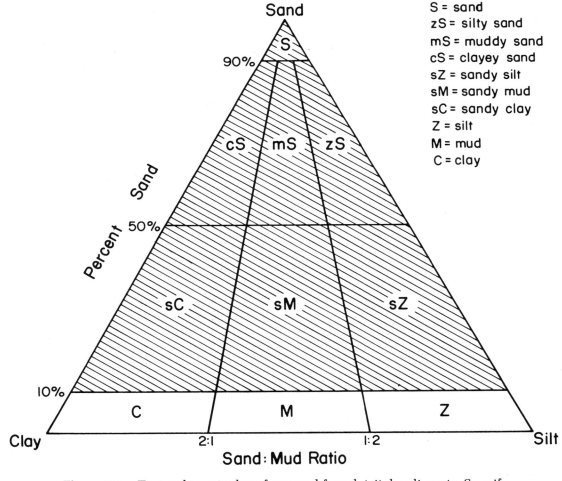

Figure 39. Textural terminology for gravel-free detrital sediments. Specify modal size of sand in crosshatched area. *(After Folk, Andrews, and Lewis, 1970)*

sophisticated than a hand lens and a grain size comparator. However, it is not possible in the field to precisely assign samples that fall close to class boundary lines. Only analysis in the laboratory can give accurate proportions for each size fraction and enable boundary-line samples to be precisely assigned or plotted samples precisely on the triangles.

Without laboratory analysis, resolution of the silt to clay ratio of the mud fraction is difficult. A qualitative field test is to grind a small sample of mud between your teeth—if silt is present you will sense it! If both silt and clay are present, qualitatively class the sample as muddy or a mud(stone). After laboratory analysis, if the sediment has between one-third to two-thirds silt or clay, use the quantitative terms "muddy (s.s.)" or "mud (s.s.)" (the "s.s." stands for *sensu stricto* and serves to indicate the specific nature of the usage).

Auxiliary Information

MODE

The most abundant size of grain *(mode)* is the measure of average size most readily determined in the field. The modal size in the gravel and sand fractions should be specified because it is valuable in determining gross depositional trends in a rock unit or sediment body and for provenance studies. Use of the mode is also valuable because it serves to distinguish among the wide range of sediment types represented in any one textural class, and it provides a far more precise impression of the size distribution characteristics. For example, both a *granular coarse sandstone* and a *cobbly fine sandstone* belong to the *gravelly sandstone* class, yet they are distinctly different and have quite different geological significance.

In specifying the modal size, the gravel and sand fractions are considered independently of each other. The modal grain diameter of each is expressed in terms of the equivalent Udden–Wentworth size grade, for example, boulder as in boulder conglomerate, or fine sand as in granular fine sand. The diameter of the modal size in the gravel fraction is always determined regardless of the proportion of gravel in the sediment. The modal size of the sand fraction is determined only for the very sandy gravel-bearing sediments (cross-hatched area of Fig. 38). It is determined for the sand fraction of all gravel-free sediments (cross-hatched area of Fig. 39).

The sand fraction in many sediments contains not one dominant grain size, but two abundant grain sizes with very few grains of intermediate size. Such bimodality provides important clues to the history of a sand. Its presence should be indicated by inserting the word *bimodal* before the term *sand* and specifying the two modes in terms of the equivalent Udden–Wentworth size grades. Thus, instead of *silty sand* (sS), the name may be, for example, *silty bimodal medium and very fine sand*, and instead of *gravelly sand* (gS), *pebbly bimodal medium and very fine sand*. This procedure should be adopted where bimodality is marked in the textural classes that are cross-hatched in Figures 38 and 39. Bimodality in the gravel or the mud fraction does not have the same significance as in the sand fraction and does not need to be specified.

SORTING

For samples falling in the two sand classes S and zS of Fig. 39, it is useful to specify the size sorting of the fraction of sand plus coarse silt. The degree of sorting is a reflection of the energy level in the environment of deposition. Three classes of sorting are readily determinable in the field: *well sorted*, *moderately sorted*, and *poorly sorted*. If the central two-thirds of the grain size range falls within less than the equivalent of one Udden–Wentworth size grade, the sample is well sorted; if it ranges over the equivalent of one to two Udden–Wentworth size grades, the sample is moderately sorted; if it ranges over more than the equivalent of two Udden–Wentworth size grades, the sample is poorly sorted. These three sorting classes are consistent with detailed classifications that sedimentologists would use in studies involving laboratory analysis.

In determining the size sorting of sand, there is one exception to the above procedure. Some bimodal sands that would be considered poorly sorted as a whole consist of two well-sorted modes. Such sands should be called "well-sorted bimodal" sand, for example, "well-sorted bimodal medium and very fine sand."

In thin section there is a common tendency to underestimate the sorting of sands, because the plane of the section does not cut through the centers of most grains. Accurate grain size analyses are generally impossible, but consistent estimations are possible with experience. Fig. 40 provides a standard for comparison to help distinguish between the better-sorted sands in thin section.

Figure 40. Comparison diagrams for visually estimating sorting in thin section. Comparative figures prepared from artificial mixtures of Pleistocene dune sand with rounded quartz grains, sieved at ¼ϕ intervals, blended to produce the distributions illustrated, then cemented into epoxy blocks that were sectioned. Silhouettes from the sections are illustrated at critical intervals for the Folk (1951) textural maturity scale (see page **68**).

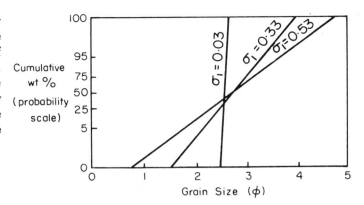

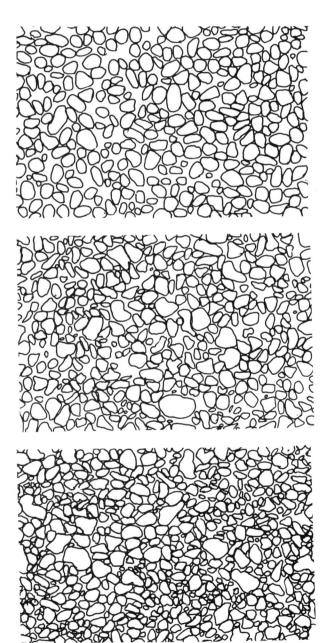

$\sigma_1 = 0.0$ very well sorted

$\sigma_1 = 0.35$

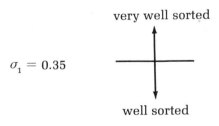

very well sorted

well sorted

$\sigma_1 = 0.5$

well sorted

moderately sorted

ROUNDNESS

The roundness of gravel and sand clasts should be noted in any rock description, although it need not be mentioned in the short name for a detrital sediment. Roundness indicates the extent of abrasion the grains have undergone. Extent of abrasion reflects overall transport history but does not necessarily reflect the distance the grains have traveled from their source— rounded grains may have been derived locally from a sedimentary rock, or may have been extensively abraded in an environment near the source, such as a beach adjacent to a cliff. In some cases, such as in soils, chemical action rounds grains (see, for example, Crook, 1968).

Quantitatively, true roundness is generally expressed by the Wadell formula:

$$\Sigma \frac{r}{R} \, N$$

where r = radius of curvature of grain corners, R = radius of largest *inscribed* circle, and N = number of corners. (For consistency, the silhouettes from which such measurements are made should be taken from grains oriented to show maximum projection sphericity —that is, viewing the plane that includes the long and intermediate axes.) Other quantitative measures of roundness exist (see Barrett, 1980).

Unless highly detailed work is justified by the likely results, practical measures of roundness rely on visual comparison with standard silhouette charts (see Fig. 41 for sand grains; see Krumbein, 1941, for pebbles).

Roundness of grains within a sediment will commonly vary even if all grains have been subjected to the same history of abrasion (or solution). Minerals and rock fragments differ in their physical and chemical properties—for example, hardness, brittleness, type of internal anisotropism, and solubility. Hence, in comparing samples it is necessary

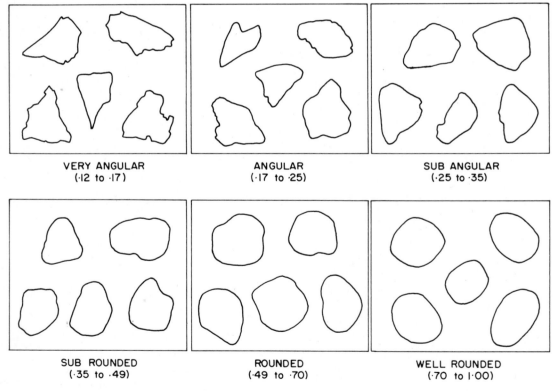

VERY ANGULAR
(·12 to ·17)

ANGULAR
(·17 to ·25)

SUB ANGULAR
(·25 to ·35)

SUB ROUNDED
(·35 to ·49)

ROUNDED
(·49 to ·70)

WELL ROUNDED
(·70 to 1·00)

Figure 41. Silhouette comparison diagram for sand grain roundness. Numbers given below the names are the quantitative limits to the classes according to the Wadell formula. (The boundary between subangular and subrounded classes corresponds to the borderline between the mature and supermature stages of textural maturity [Folk, 1951].) (*After Shepard and Young, 1961)*

to contrast roundness of the same type of component. Quartz is most commonly used in sandstones because it is abundant, hard, and has relatively isotropic physical and chemical properties. In addition, grains of different size round at different rates—the coarser the faster. Furthermore, even if the grain assemblage as a whole has the same history, individual grains within the assemblage will have been subjected to somewhat different degrees of abrasion or solution (for example, in high-energy environments some grains may be fractured by collisions or by being ground between larger grains to produce "broken rounds"). It is generally necessary to record the *range* in roundness as well as the average, and to note any differences correlated with size or composition.

The two most important characteristics to look for are *trends* of changing roundness (vertically or laterally in a succession) and *unexpected differences* (such as marked differences in roundness between grains of the same size and composition; finer grains with higher degree of roundness than coarser grains of the same composition; harder, more brittle, or more anisotropic grains with higher degree of roundness than softer, more ductile, or more isotropic grains).

SPHERICITY

Sphericity is a measure of how nearly equal the axial dimensions of a particle are. *True sphericity* is the surface area of a grain divided into the surface area of a sphere of the same volume—a rather impractical property to measure! Operational sphericity (Wadell, 1935) is:

$$\sqrt[3]{\frac{Vp}{Vcs}}$$

where Vp = volume of particle and Vcs = volume of smallest sphere that would enclose the particle. Vcs is approximated (Krumbein, 1941) by

$$\sqrt[3]{\frac{LIS}{L^3}} = \sqrt[3]{\frac{IS}{L^2}}$$

where I = intermediate axis, S = short axis, and L = long axis. Other measures exist (for example, see Barrett, 1980). Most quantitative studies plot various functions of axial lengths on graphs (for examples, Fig 42). For qualitative studies, silhouette charts are available for visual comparison (for example, Rittenhouse, 1943).

Sphericities of gravel clasts may warrant detailed study. Sphericity is strongly influenced by physical anisotropism of the rock and mineral particles; hence comparisons must be between clasts with the same structure, texture, and composition. Particular clast shapes seem to concentrate in particular environments (for example, discs on beaches, rollers and blades in rivers) either because of selective sorting or because of actual modification to grain shape by environmental processes. However, to date insufficient work has been performed to provide firm interpretive generalizations, and spher-

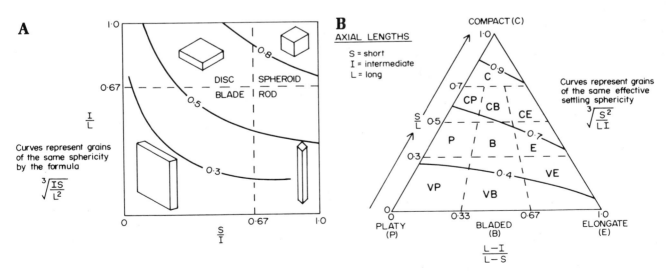

Figure 42. **Two systems for representing particle shape.** *A:* Shape classification after Zingg (1935, in Pettijohn, 1975). The representative solids have the same minimal roundness values. *B:* Sphericity-form diagram after Sneed and Folk (1958)

icity studies should be integrated with other data in any individual case.

Sphericity of sand grains is generally not reported unless there is a significant population of inequant particles. It cannot be reliably estimated in thin section, but hand specimens of indurated rocks should be examined to determine whether a preferred fabric has been imparted by inequant grains; if so, "size" measurements in thin section may be strongly biased. Fabric imparted by inequant grains may be useful in paleocurrent studies (for example, imbrication and current lineation). Detailed sphericity analysis of loose sand grains also can prove useful in determination of paleohydrodynamic conditions (see Moss, 1972), particularly when combined with studies of size distributions and sedimentary structures.

SURFACE TEXTURE

Detailed study of the surface features of detrital sand grains requires use of the scanning electron microscope (SEM). Different polishes or surface pits and protrusions appear to be caused by different processes, and these processes may be characteristic of certain environments. Hence the surface textures may be used to infer the history of grains or grain assemblages (for example, Krinsley and Doornkamp, 1973; Baker, 1976). There are many limitations to these studies, however, and results are often equivocal. Representative sampling is a problem.

Surface textures of gravels, determined by visual inspection, can be useful. Glacial processes (and gravitational mass movements) may produce striations or facets; wind may sandblast pebbles into einkanter or dreikanter and etch surfaces; beach reworking may produce chink facets; fluvial action may produce percussion marks and spalls; and dif-

ferential chemical solution (sometimes assisted by pressure at grain contacts) can produce pits.

FABRIC

Packing and orientation of grains can be studied relatively easily in gravel deposits, although problems arise with some well-indurated conglomerates. For interpretive purposes, it is important to note whether the framework clasts are clast- or matrix-supported, have a preferred orientation (for example, of long axes, or of the planes containing the intermediate and long axes), are imbricated, or show grading (normal, inverse, or symmetrical). Beware of deformation fabrics produced *after* deposition by an external stress (even compaction in some cases may produce a preferred orientation).

Fabric of sand grains has been little studied because of the difficulties involved in sampling loose modern sediments and in determining any fabric in three dimensions. However, some useful orientation studies have been made in sedimentary rocks (see Pettijohn, 1975), and any obvious fabric should be noted not only because it could prove informative, but because it may affect size analysis. Packing of grains (see Fig. 43) controls the way in which sediments respond to stress (for example, thixotropy and dilatancy), porosity and permeability (see Halley, 1978; Curtis in Carver, 1971), and the apparent size distribution of sediments in thin section.

Fabric of the clayey components in sedimentary rocks also has been little studied by petrologists. Soil scientists have given more attention to this attribute, and the descriptive terminology given in Table 5 may be worth applying in detailed geological studies. This terminology is only part of the much broader system presented by Brewer (1964) —modifications or additions to his schema may be warranted. Thin sections must be used.

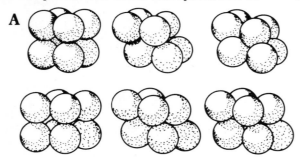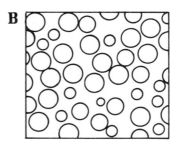

Figure 43. *A:* Ideal packing configurations for equal-size spheres. Upper left is cubic ("open") packing; lower right is rhombohedral ("closest") packing. *(After Graton and Fraser, 1935)* *B:* Random cut through open-packed arrangement of equal-size spheres.

Table 5. Terminology Used to Denote Clayey Fabric Distribution and
Orientation Patterns

Fabric Distribution Patterns

Overall	*Individual Particles*	*Relation to Reference Feature*
porphyroskelic (dense groundmass in which grains are set)	random	unrelated
	clustered	normal (perpendicular to . . .)
	banded	
intertexic (grains touching or linked)		parallel
	radial	
	concentric	inclined (state angle to . . .)
granular (no true matrix)		cutanic (concentrations forming a unit mass adjacent to . . .)

Fabric Orientation Patterns (note magnification used)

Individual Particles

Oriented: Individuals sufficiently oriented that there is roughly continuous birefringence Under Crossed Nicols (UXN); with rotation of the stage, dark extinction lines or bands move across the aggregation of individuals. Can be subdivided into strongly, moderately, or weakly oriented, although attitude of thin section relative to fabric may influence apparent degree of orientation.

Unresolved: At magnification used, appears to be some anisotropism within aggregates that are too small to observe in detail; UXN the mass appears to have random fabric.

Unoriented: At magnification used, fabric is isotropic due to apparent random orientation of individual particles.

Indeterminate: Fabric appears isotropic because of opacity or crystallographic character.

Domains (aggregations of clayey material, each aggregation with a preferred orientation

Asepic fabric: Dominantly anisotropic, with anisotropic domains unoriented with respect to each other.

Sepic fabric: Various recognizable anisotropic domains with various patterns of preferred orientation (see Brewer 1964, for subdivisions).

Undulic fabric: Practically isotropic at low magnifications, and weakly anisotrophic with faint undulose extinction at high magnifications. Domains not distinct.

Isotic fabric: Apparently isotropic matrix at all magnifications.

Crystic fabric: Anisotropic fabric involving recognizable crystals deposited during pedogenesis or diagenesis (see Brewer, 1964, for subdivisions).

Strial fabric: Clayey material as a whole exhibits preferred parallel orientation, giving extinction pattern that is either unidirectional or shows several preferred extinction directions. Common in sedimentary rocks, probably imparted by compaction and other diagenetic processes. May be superimposed on other fabrics, which should be sought when strial fabric is at extinction.

Source: after Brewer, 1964

QUALITATIVE INTERPRETATION OF TEXTURES

Textural properties of a detrital sediment provide guidelines to interpretation even if you do not perform rigorous quantitative analysis (see following section). The preceding section mentioned some of the ways you can use them; this section offers generalizations about the significance of size distributions and roundness.

The size distribution in a sediment mainly reflects the conditions in the depositional environment—processes acting and the energy level of those processes. For example, sandy beaches are almost always composed of well-sorted, matrix-free sands because waves exert energy continuously for prolonged periods on these sediments. Particles are segregated according to their hydrodynamic behavior, which largely depends on their size (although specific gravity and shape are also influential). Fines are winnowed away, and coarse grains are either concentrated in the highest energy locale or buried. Most sandy floodplain deposits, in contrast, tend to be less well sorted and to contain some muddy matrix because river energy levels fluctuate and deposits are not continuously reworked. Deposits from turbidity currents tend to be poorly sorted and contain abundant matrix because of rapid sedimentation under low energy conditions. When the competency and capacity of currents are high, the coarsest grains that can be transported travel with all the finer particles; if capacity and competency are rapidly lost (that is, if the driving force of the current wanes abruptly), all the sizes will accumulate together. On the other hand, if the capacity and competency wane gradually, grains of only one size will accumulate at one time at any one locality—the finer particles will be carried on. Hence, deposits of floodplains may be well sorted locally, and sorting improves upward in turbidites. Of course, size distributions may also be influenced by the previous history of the sediments—for example, eolian deflation of the fine particles in an earlier stage of the sediment history may leave only coarse grains for transport by a high energy process. Or, if no gravels exist in the area of sediment supply, of course no gravel can be transported or deposited regardless of the competency of a current. In addition, postdepositional processes may modify the size distribution of a deposit—examples are the biogenic mixing of initially well-sorted layers (see Faas and Nittrouer, 1976), the pedogenetic infiltration of colloids and clays (see Brewer and Haldane, 1957), and the production of clayey matrix by

diagenetic breakdown of unstable minerals (see Whetton and Hawkins, 1970).

Fig. 44 presents graphs that indicate some general relationships between sediment sizes and their behavior in water. In conjunction with information derived from other textural attributes (such as shape and roundness), compositional attributes (such as the range of specific gravities of the particles), sedimentary structures, and the vertical (and lateral) stratigraphical relationships, these provide a qualitative guideline for interpreting depositional conditions. They cannot be used in a rigorous quantitative sense because too many influential variables are generally unknown (for instance, local availability of particle sizes, volume and depth of the transporting current, degree of turbulence, frictional resistance to flow of the substrate, and total quantity of sediment that was being transported). Note that clay sizes settle so slowly that *any* current with an upward component would prevent deposition—most clays probably aggregate as fecal pellets (see Haven and Morales-Alamo, 1968, or Prokopovich, 1969) or flocs (for example, see Pryor and Vanwie, 1971) and accumulate as relatively coarse particles.

Textural Maturity

Folk (1951) devised a scale of "textural maturity" for detrital sandstones which represents a qualitative guideline to the "expected" progression of several textural attributes as progressively more energy is expended on a sediment:

Immature stage: sediment with more than 5% detrital clays; sand grains poorly sorted and angular.

Submature stage: sediment with less than 5% detrital clays; sand grains poorly sorted and angular.

Mature stage: sediment with less than 5% detrital clays; sand grains well sorted but still angular. (See Fig. 40 for the critical boundary between "moderately" and "well" sorted grains in thin section views.)

Supermature stage: sediment with less than 5% detrital clays (essentially no clay); sand grains well sorted and rounded. (See Fig. 41 for the critical boundary for "rounded" grains in this scheme.)

The stage attained by any sandstone provides a basis for initial interpretation of the sediment history. So-called *textural inversions* or deviations from the listed association of textural parameters are common. Recognition of such deviations is im-

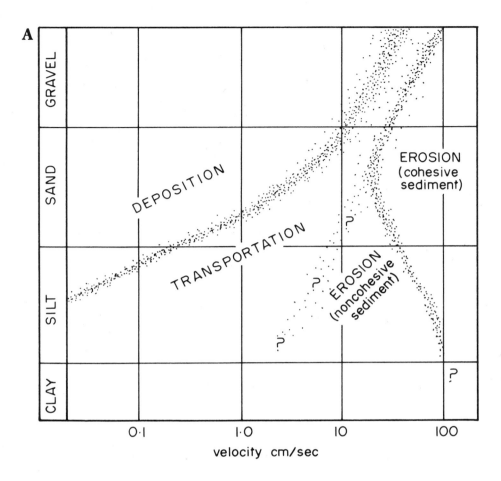

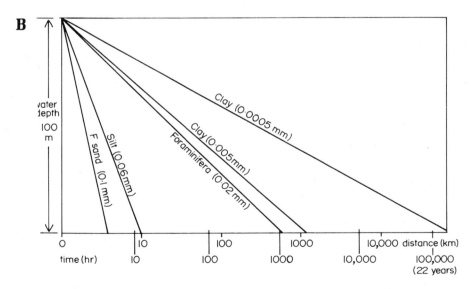

Figure 44. Diagrams illustrating behavior of sediments in water. *A*: Schematic Hjulstrom diagram, showing approximate relationships between current velocities and behavior of stream-bed sediments. Few reliable data exist for particles finer than 0.2 mm.

(After Heezen and Hollister, 1964)   *B*: Settling times and distances of selected sediment sizes, assuming horizontal current flow of 10 cm/sec. *(After Garrels and Mackenzie, 1971)*

portant because they imply that something unusual has happened and require you to think about and search for the cause of the inversion. Examples are rounded grains in poorly sorted deposits (possibly due to mixing of originally well-sorted layers after initial deposition by storms or bioturbation, or possibly a supply of rounded grains from a sedimentary source rock) and well-sorted grains with a clayey matrix (perhaps finally deposited after an energetic history in a very low energy environment, such as sand dunes or a barrier bar migrating into a lagoon, or perhaps due to the pedogenetic infiltration of clays). Other inversions (such as bimodal grain roundness) and other explanations for them can be imagined. As a rule, you should class the sediment according to the lowest stage of textural maturity that it shows (thought to be indicative of the last set of processes acting on it), qualified by the kind of textural inversion you note. The likely cause(s) for the inversion will generally require extrapolating from your knowledge of other attributes of the deposit or the stratigraphic succession. Figure 45 illustrates the range of textural maturity that is most commonly represented in various environments.

QUANTITATIVE ANALYSIS OF TEXTURES

Methodology

Quantitative analysis of the size distributions of detrital sediments is necessary for detailed comparison between samples and to discover signifi-

cant relationships between sediment properties and geologic processes or settings (see, for example, Krumbein, 1968; Klovan, 1966). Often subtle differences, obscured by a purely qualitative description, are significant in establishing trends and for interpreting depositional conditions. Methods of quantitative analysis emphasize practicality—procedures are simple, rapid, and precise (that is, they provide reproducible results). The concept of accuracy, or approach to a measure of "true size," is not involved unless comparison is attempted between samples analyzed by different methods (for example, see Griffiths, 1967). (In fact, the concept of "size" cannot be completely separated from that of "shape.") Each method measures a different attribute of "size", and in none of the widely used methods is there a close relationship between measuring techniques and natural transportational processes. Hence it is not particularly important which measurement technique you adopt, as long as you analyze all samples to be compared in precisely the same way. Standardization of procedure is crucial; comparison of size distributions obtained by different methods is fraught with problems!

Folk (1966), Moiola and Weiser (1968), Jones (1970), and Tucker and Vasher (1980) review various methods of statistical analysis. If you are to perform an analysis professionally, select the most suitable method for your purpose and when publishing, inform your audience which one you used.

One of the most important things to do before you begin quantitative analysis is to ensure that you have obtained *representative* samples from the sediments. In each case, you should sample a single sedimentation unit (deposited under essentially

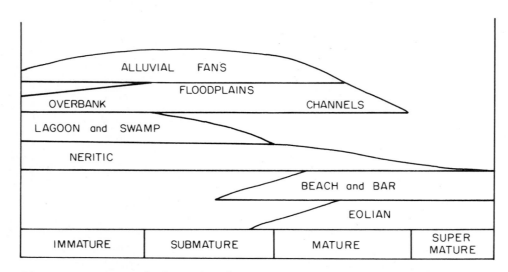

Figure 45.   Expected relationships between environments and textural maturity.

constant physical conditions); this requirement may mean sampling individual laminations from some lithological units (see, for example, Emery, 1978; Grace et al., 1978; Macpherson and Lewis 1978). In the laboratory, you must take equal care to obtain representative subsamples (see Krumbein and Pettijohn, 1938; Cochran, 1977).

Before analysis, samples must be fully disaggregated and dispersed. Soluble salts and grain coatings may have to be removed (see separate section on "Pretreatment of Samples"). Where diagenetic modification to grains has taken place, size analysis is generally not warranted.

Most data on the sand (and gravel) fraction have been, and are, obtained from sieve analysis, which sorts on the basis of smallest cross-sectional diameter. Since most grains are not spherical, the number of particles that pass through a given sieve is time-dependent—there will always be more grains that *could* pass through a given sieve if they were to land with just the right orientation on the mesh. Hence the time of sieving must be standardized, as well as other procedures. A comprehensive "cookbook recipe" is provided later. Slight variations may be necessary for particular sample suites or may be suggested for use in your particular laboratory; if so, make sure you follow the same procedure with all samples.

A growing number of laboratories are using settling tubes for sand analysis (for example, see Reed et al., 1975; Taira and Scholle, 1979). The procedure involves using smaller samples (say, 1-2 g) than those used in sieving, and special care must be taken to obtain representative subsamples from the initial sample. A great many instruments are available, with various diameters and lengths of tube, and various devices for measuring the proportion of grains that reach a given point or accumulate on the bottom in relation to time. Each laboratory details specific analytical procedures for its particular instrument.

Most data on the mud fraction are obtained from pipette analysis. The principle is the same as that for settling tube analysis of sand—sizes are calculated on the basis of an ideal settling velocity (v) formula, usually Stokes' Law: $v = Cd^2$, where d = the diameter in centimeters of an assumed sphere and $C = (ds - df)g/18\mu$ in which ds = density of solid (quartz = 2.65g/cm^2); df = the density of the fluid you are using at the particular temperature; g = the acceleration of gravity (980cm/sec^2); and μ = the viscosity of the fluid at the particular temperature.) A standard procedure is described later. Electronic instruments are also widely used for mud size anal-

ysis—again, many types are available and specific procedures vary. They appear to produce coarser, better sorted results than pipette analysis because they do not evaluate grains finer than 11 $\varnothing$ (see Kelley, 1981).

Hydrometer methods for analyzing the mud fraction are also widely used, especially by soil scientists. Direct measures of the decreasing density of a suspension with time are related to the size distribution. Experiments in the author's laboratory suggest that results are comparable to those from pipette analysis, thus a detailed standard procedure is also given for this method.

Treatment of Data

Once you have obtained raw data on the "size" distributions of samples, you can use various means to obtain summary statistical parameters that represent characteristics of these distributions. The parameters are mathematical artifacts—numbers computed by standard formulas—and represent only a few characteristics of each distribution. Graphs are necessary to represent entire distributions (see Fig. 46). Note that histograms and frequency curves derived from histograms are misleading; cumulative frequency curves are generally used, plotted on probability paper when accurate interpolation from the "tails" is necessary. To obtain the statistical parameters, you can use either the entire body of data (the purely mathematical "Method of Moments"), or you can take selected values from the cumulative curve and insert them into formulas (results only approximate the values obtained from the Method of Moments). Note that in all calculations, *weight* % is used rather than frequency %. Statistical formulas are designed fundamentally for the latter, but counting of grains is impractical and we *assume* there is no significant difference in grain densities between samples we compare—an assumption that is invalid if comparisons are attempted between quartz-rich sediments and blacksands, greensands, or some rock-fragment- (or shell-) rich sands.

The *Method of Moments*, which is widely used in this age of computers (see Friedman, 1979), has marked advantages over graphical analysis in speed and freedom from some bias inherent in the assumption of a fundamental lognormal size distribution (see Leroy, 1981). Use of the method requires the *complete* size distribution of each sample—there can be no "open tail" (that is, no unsized "pan fraction"; data must be extrapolated to 100% at an arbi-

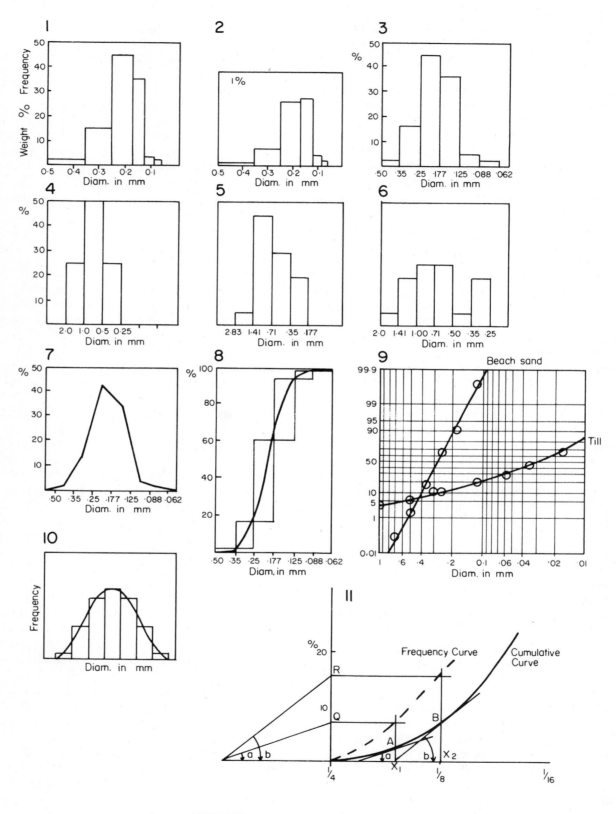

$$\emptyset \ = \ - \log_2 \ (\text{diameter in mm})$$

Figure 46 (at left).   Graphic presentation of size data.

1. The Abscissa scale is *arithmetic,* and therefore bars of ½φ intervals span different widths. Note that with sedimentary geological studies, the coarse size is to the left on the abscissa.

2. Diagram 1 plotted correctly for bar graphs, on the basis of unit *areas.*

3. The abscissa scale is *geometric* (although it is expressed in millimeter values). Data are the same as those in diagram 1 with one size class added. Note that the modal class *appears* different from that in 1.

4, 5, 6. Data are the same. Sketches 4 and 5 use an abscissa scale with 1φ intervals, starting at different points (−1φ for 4, −1.33φ for 5); 6 is subdivided at ½φ intervals. Note the very different shapes that result from varying the abscissa scale.

7. This frequency polygon graph was constructed by joining the midpoints of the bars shown in 3. The graph is subject to all the same problems as demonstrated in the bar graphs—the shape will vary depending on the abscissa scale.

8. A cumulative curve can be constructed using bar graph data (of diagram 3). The shape of the curve will be independent of the abscissa scale. In practice, raw data are used directly to define points for the curve.

9. Cumulative curves may be drawn using probability ordinates. Because of the standard usage of the Udden-Wentworth grade scale, either an arithmetic abscissa (with φ values) or a log₂ abscissa scale (when millimeter values are plotted, as shown) can be used. If the limits of size measurement are attained before the data line reaches 100% (as in the case of the till figured), it may be necessary to make an arbitrary assumption that the fine tail has a normal distribution and a minimum size of 14φ—permitting a straight-line extrapolation from the last data point to 100% at 14φ.

10. A frequency curve derived from a histogram. Such graphs will have the same problems as histograms, varying in shape depending on the abscissa scale used. They are *more* misleading than histograms because one *thinks* one can select a unique modal size.

11. Construction of the unique frequency curve. A cumulative curve is necessary, then a complex graphical technique must be employed (see Krumbein and Pettijohn, 1938).

trarily selected size at the fine end if there is a fraction too fine for practical measurement). Size values used in the computations assume a "normal" (Gaussian) distribution within each class that has been measured (in the example below, the value of *d* is midway in each φ class). This assumption is generally false and can distort (increase) standard deviation results. The mode (most common grain size) cannot be determined, nor can any polymodality in the distribution. Table 6 shows an example of Moment calculations.

Graphical analysis is time-consuming because not only must curves be drawn but also values must be interpolated from them. However, computers can be programmed to do both. The graphical statistical parameters of Folk and Ward (1957; Table 7) are most widely used. Although results only approximate the Moment values, there are advantages in their use when samples have "open tails" or when there is likely to be experimental error in measuring the size of the coarsest or finest grains—5 to 8% of the distribution at each end is ignored. Plotting of curves has other advantages: sample interrelationships—differences and similarities—can be seen throughout their range; modes can be found and polymodality seen; defective sieves may be detected ("kicks" at the same φ value in a bundle of curves); qualitative textural classification of the sample, as discussed previously, can be determined readily by visual inspection.

Statistical parameters can be easily and rapidly estimated after only slight experience with graphs. The mode(s) occurs at the inflection point of the steep part(s) of the curve (most easily found on graphs with arithmetic ordinates). You can find the median of the *sand* fraction (versus the median of the whole sample, which is merely the φ value at 50 wt. %) by taking the average of the percentage values at −1 and +4φ. To obtain an estimate of the standard deviation consistent with the qualitative estimate of "sorting" made from looking at the actual sample, look at the central two-thirds of the curve (that part of the curve between 16 and 84%). If it spans less than 1 φ unit, the sample is well sorted; if it spans 1 to 2 φ units, the sample is moderately sorted; if it spans more than 2 φ units, the sample is poorly sorted. For skewness (a measure of symmetry of the distribution), sketch a line between the points on the curve at 16% and 84%. If the line lies very close to the (overall) median, the distribution is near-symmetrical; if the line lies to the right of the median, it is fine-skewed (+ *skewness*); if it lies to the left of the median, it is coarse-skewed (− *skewness*). For kurtosis (a measure of

Table 6. Example of Method of Moments Calculation

The general formula for the nth moment is

$$\log n = \frac{\Sigma(fd^n)}{N}$$

where f = frequency (weight %), *d* = log diameter, and *N* = number of measurements (100 when dealing with percents.)

Given: an ideal "normal" frequency distribution:

ϕ	d	f	fd	d^2	fd^2	d^3 fd^3 d^4 fd^4
−4 to −3	−3.5	3	−10.5	12.25	36.75	
−3 to −2	−2.5	10	−25.0	6.25	62.5	
−2 to −1	−1.5	22	−33.0	2.25	49.5	unlimber
−1 to 0	−0.5	30	−15.0	0.25	7.5	the
0 to +1	+0.5	22	+11.0	0.25	2.75	calculator!
+1 to +2	+1.5	10	+15.0	2.25	22.5	
+2 to +3	+2.5	3	+07.5	6.25	18.75	
		$N = 100$	$\Sigma fd = -50$		$\Sigma fd^2 = 200.15$	

The first moment is the *mean:*

$$\frac{\Sigma fd}{N} = \frac{-50}{100} = \underline{\underline{-0.50\phi}} = \overline{X}$$

This is a log mean, hence the ϕ value; the antilog$_2$ gives the millimeter value.

The second moment is the variance:

$$\frac{\Sigma fd^2 - (\Sigma fd)^2}{N_1} = \frac{200.15 - 25}{99} = \underline{\underline{1.77\phi}}$$

The standard deviation is

$$\delta = \sqrt{variance} = \sqrt{1.77} = \underline{\underline{1.33\phi}}$$

The third moment will give a value interpretable as skewness, and the fourth moment a measure of kurtosis:

$$3rd\ moment - skewness = \frac{\Sigma f(d - \overline{X})^3}{100\delta^3}$$

$$4th\ moment - kurtosis = \frac{\Sigma f(d - \overline{X})^4}{100\delta^4}$$

the peakedness of the distribution), the spread between 5 and 95% is 2.44 times the spread between 25 and 75% for a *normal curve (mesokurtic);* if the ratio is much less than 2.44, the curve is *platykurtic* with a K_G less than 1.00; if the ratio is greater than 2.44, the curve is *leptokurtic* (excessively peaked) with a K_G greater than 1.00. Try your hand with Fig. 47.

Interpretation

For any *interpretation* from quantitative size analysis, a suite of samples must generally be examined—little of value will result from studying merely one or two samples. Apart from their use as sediment descriptors and for discovering similarities, differences, or trends in sample suites, the

Table 7. Formulas and Verbal Scales for Folk and Ward (1957)
Sedimentary Grain Size Parameters

Mode: Most frequently occurring particle size. Inflection point(s) on the
steep part(s) of the cumulative curve. Found precisely only by trial
and error method: discover the point with the maximum wt% within
a ½ ϕ interval centered on it.

Graphic Mean:
$$M_z = \frac{\phi 16 + \phi 50 + \phi 84}{3}$$

Inclusive Graphic Standard Deviation:
$$_1 = \frac{\phi 84 - \phi 16}{4} + \frac{\phi 95 - \phi 5}{6.6}$$

<0.35ϕ	very well sorted
0.35 to 0.50ϕ	well sorted
0.50 to 0.71ϕ	moderately well sorted
0.71 to 1.0ϕ	moderately sorted
1.0 to 2.0ϕ	poorly sorted
2.0 to 4.0ϕ	very poorly sorted
>4.0ϕ	extremely poorly sorted

Inclusive Graphic Skewness:
$$Sk_1 = \frac{\phi 16 + \phi 84 - 2\phi 50}{2(\phi 84 - \phi 16)} + \frac{\phi 5 + \phi 95 - 2\phi 50}{2(\phi 95 - \phi 5)}$$

$$= \frac{\phi 84 - \phi 50}{\phi 84 - \phi 16} - \frac{\phi 50 - \phi 5}{\phi 95 - \phi 5} \quad \text{(see Warren, 1974)}$$

+1.0 to +0.3	very fine-skewed
+0.3 to +0.1	fine-skewed
+0.1 to −0.1	near-symmetrical
−0.1 to −0.3	coarse-skewed
−0.3 to −1.0	very coarse-skewed

Graphic Kurtosis:
$$K_G = \frac{\phi 95 - \phi 5}{2.44(\phi 75 - \phi 25)}$$

<0.67	very platykurtic
0.67 to 0.90	platykurtic
0.90 to 1.11	mesokurtic
1.11 to 1.50	leptokurtic
1.50 to 3.00	very leptokurtic
>3.00	extremely leptokurtic

results of statistical analysis have generally been applied to discovering depositional environments of "unknowns" by comparing their size distribution characteristics with a data base obtained from analyzing samples from "known" environments. Much sedimentology has been, and still is, directed to establishing this data base. Whereas some generalizations can be made about distributions that are characteristic of some environments (such as beaches, which are generally well sorted), a variety

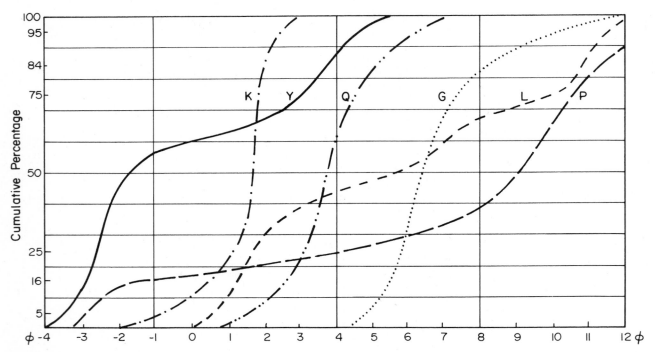

Figure 47. Cumulative curves for textural interpretation. For each curve, determine the grain-size name (e.g., pebbly, silty, very fine sand) and estimate dispersion (Folk Verbal Scale), skewness (coarse-skewed, near-symmetrical, fine-skewed), kurtosis (leptokurtic, mesokurtic, platykurtic), and mode(s).

of processes act in most environments at varied energy levels—and it is the processes and their energy levels that dictate size distribution characteristics. (See Allen, 1971, for an excellent example.) Hence, modern studies are concentrating on interpreting the *process* of transportation/deposition, with the ultimate intention of being able to infer paleoenvironments from a knowledge of the kinds of processes that characterize them.

Fundamental studies of the relationship between grain size (and shape) distributions and processes are as yet few and have shown that there are great complexities (see, for example, Tanner, 1964; Davis and Erlich, 1970; Moss, 1972; Middleton, 1976; Middleton and Southard, 1978; Erlich et al., 1980). Grain size depends on the character of the source rocks, weathering processes, abrasion, and selective sorting during transportation. During transport, grains travel by three different mechanisms: *traction* (continuously in contact with the substrate), *intermittent suspension* (bouncing along the substrate), and *suspension* (in the transporting medium above the substrate). The proportion of grains traveling in each way depends on the competency and capacity of the transporting agent (reflecting factors such as volume, velocity, bed roughness, and turbulence), on the range of grain sizes avail-

able (reflecting ultimate and local source factors— no gravels will be transported if none are locally available, however strong the current), and on boundary conditions between the different transporting mechanisms (particularly the transition from traction to intermittent suspension). One method of interpreting process from grain size distributions attempts a simple analysis of the complete cumulative curve, separating populations that appear to comprise straight-line segments of the curve (see, for example, Visher, 1969, and Fig. 48). Insofar as these "populations" are artifacts of the method of analysis, interpretations still depend on comparing characteristics of "unknowns" with a data base of "knowns." Another approach is to dissect the frequency curves into assumed lognormal populations (see Tanner, 1964) and to analyze each of these separately (see Middleton and Southard, 1978, and Fig. 48). This method has more promise for discovering relationships between processes and deposits. However, original sedimentary grain populations almost certainly do not have lognormal distributions, and it remains to be seen whether the false assumptions made about the ideal distributions will invalidate these approaches. (They may not—if the results do prove useful, there is every reason to use them!)

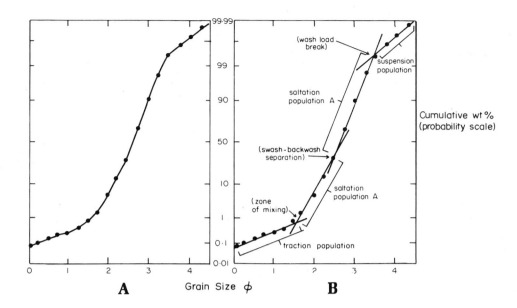

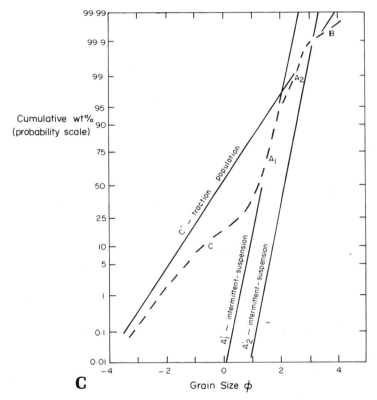

Figure 48. Examples of graphical dissection in statistical analysis of grain-size data. A and B: Simple graph dissection method of Visher (1969. A: Cumulative curve of size distribution. B: Dissection of cumulative curve into four truncated log-normal populations from a beach foreshore sample (and interpretations). Straight-line segments are thought to reflect transportational processes. C: Fitting of overlapping lognormally distributed populations (solid lines) to observed distribution (dashed line). *(After Dalrymple, in Middleton and Southard, 1978)* Four populations are suggested by the 3 inflection points in the original curve (C, A$_1$, A$_2$, B). A computer program can approximate the distribution of each ideal population (C′, A$_1$′, A$_2$′, and B′ not shown), the combination of which gives the original curve. Each population is inferred to reflect a transportational process. Not all curves will show as many populations.

Yet another approach looks at trends within a suite of sediments related to a single system of environments (see McLaren, 1981). With selective (that is, incomplete) erosion from any source material, the transported sediment will be finer, better sorted, and more negatively skewed than the source, and the residual (lag) deposit coarser, better sorted, and more positively skewed. With selective deposition from any sediment distribution, the deposit may be either finer or coarser than the source material, but sorting will be better and the skewness more positive. Hence any trends that can be determined in any of the relevant statistical parameters can help distinguish source materials from deposits, determine net transport paths, indicate processes acting, and identify depositional environments.

Plots of various statistical parameters against each other can depict trends or clustering of samples that can be used in other ways (see, for example, Cronan, 1972; Folk and Ward, 1957; McCammon, 1976; Passega, 1964, 1972, 1977; and Fig. 49).

Thin-Section Analysis

The texture of rocks that cannot be disaggregated into their constituent grains must be studied in thin section. Severe limitations are inherent in such textural studies because a thin section provides an essentially random two-dimensional, planar cross-section of the rock. Hand specimens or several thin sections cut at right angles should be examined for sphericity and packing. Because thin sections cut very few grains in the plane of their short and intermediate axes (the dimensions that determine their sieve "size"), quantitative grain size data cannot be directly compared with data compiled from sieve (or setting tube) analysis. Quantitative intercomparisons of thin-section grain size data are feasible, particularly when data are compiled by the same operator (to avoid operator bias in measurement procedures) and from rocks that do not differ greatly in packing or grain sphericity (which can influence apparent size distributions). Techniques for quantitative thin-section analysis, including modal analy-

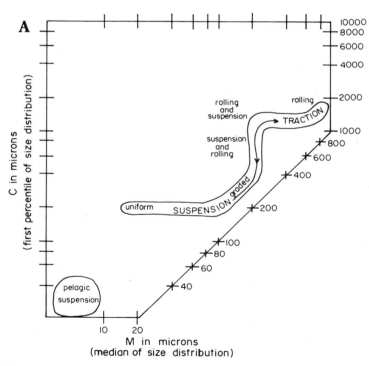

Figure 49. Examples of scatter plots using statistical grain-size data. *A*: CM diagram—a guide to depositional process. *(After Passega, 1972)* *B* and *C* are examples of scatter plots of selected Folk and Ward statistical parameters. *(After Andrews and van der Lingen, 1969)* *B*: Standard deviation versus mean size. Unimodal samples are open circles, polymodal samples are ×5. A sinusoidal field for unimodal samples appears to be a widespread phenomenon. *C*: Skewness versus standard deviation with only unimodal samples plotted. Fields enclose samples from the same locality.

sis for compositional data, are provided in Krumbein (1935), Carroll (1941), Chayes (1956), Carver (1971); also see Griffiths (1967).

Several attempts have been made to relate size distributions determined from thin sections to those from sieve analysis (see, for example, Rosenfeld et al., 1953; Friedman, 1958; Kellerhals et al., 1975; Adams, 1977). The most convincing relationship, determined by Harrell and Eriksson (1979), is outlined below (Table 8). They analyzed 84 sandstones both in thin section and by sieving; the sandstones were quartz-rich and more testing is required to determine whether their conversion factors are valid for sandstones with other common components. Almost certainly no conversion factor is possible for rocks with common inequant grains (such as many rock fragments).

Measure the apparent long dimension of 200 to 500 grains, and record the data in sets at $\frac{1}{2}\varnothing$ or $\frac{1}{4}\varnothing$ intervals. Plot cumulative curves on probability graph paper (arithmetic abscissa), with each point on the fine side of the class interval; connect points with straight lines and make no extrapolations. Apply conversion formula (Table 8).

Table 8. Harrell and Eriksson (1979) Method for Conversion of Thin Section to Sieve Size Data

General formula:
(sieve data) $= a + b \times$ (thin-section data)

Graphic Parameters	a	b	
M_d (median)	0.121	1.030	accurate
M_z	0.227	0.973	accurate
σ_1	−0.029	1.015	reasonable
Sk_1	0.049	0.593	approximate
K_G	0.394	0.706	not significant at 95% confidence level

ϕ Cumulative Percentiles	a	b	ϕ Cumulative Percentiles	a	b
2	0.164	1.137	50	0.121	1.030
5	0.156	1.097	64	0.254	0.969
9	0.178	1.066	75	0.325	0.952
16	0.127	1.075	84	0.452	0.895
25	0.117	1.064	91	0.579	0.846
36	0.094	1.054	95	0.772	0.801
			98	0.989	0.770

Note: All but ϕ91–98 appear to be accurate; ϕ98 is the least accurate but still a good approximation.

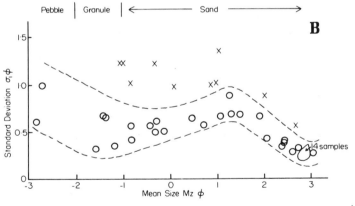

B

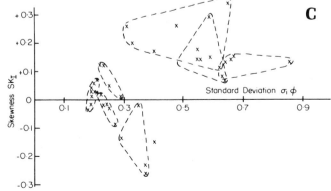

C

Figure 49. *(Continued)*

LABORATORY METHODS

Pretreatment of Samples

Depending on the sample, you may need to apply any one or a combination of the following techniques to create a loose assemblage of grains, which you may then subject to various methods of size, shape, and mineralogical analysis. Inspection with a binocular microscope (and possibly examination of thin sections) will be necessary to evaluate special problems. The results of the disaggregation can be accomplished only with a concomitant loss of the ability to analyze certain properties (for example, even gentle disaggregation techniques may modify the size and shape of soft or brittle grains, such as shell fragments). In general, take precautions not to lose (or add) any material (wash your fingers, pestles, and so on). Sieve or decant at various stages in the procedure to separate fractions requiring further or different treatment. All samples should be subjected to the same pretreatment(s) in any suite of analyses.

PHYSICAL DISAGGREGATION

1. Crush with (rubber-gloved) fingers, either dry on glazed paper or with distilled water (and dispersant) in a basin.

2. Crush gently with a rubber bung under the same conditions.

3. Stir gently in (warm) distilled water with dispersant.

4. Shake in a bottle with (warm) distilled water and dispersant.

5. Rub with a stiff brush in distilled water with dispersant.

6. Boil in water with dispersant. "Quaternary 0" (a strong detergent) is a 20% liquid solution which may be particularly effective when added to the boiling water (Zingula, 1968). Boil it for up to 1 hr, stirring occasionally.

7. Place the sample plus water and dispersant in an ultrasonic (over 20,000 cps) shaker (see Moston and Johnson, 1964; Kravitz, 1966). An ultrasonic probe device may be applied directly to resistant fragments (be sure to use a plastic beaker). (See Savage, 1969.)

8. Pound with a rubber pestle in a porcelain mortar (in water-plus-dispersant).

9. Use a porcelain mortar and pestle to pound the sample (remove fines periodically to avoid excess crushing). Avoid rotary grinding, which crushes grains.

10. Dry small aggregates of the sample thoroughly in an oven. Saturate the sample in kerosene (or gasoline). Decant the fluid. Cover the wet sample with water. The force generated by water molecules displacing the hydrocarbons may be sufficient to disintegrate mudrocks. A small amount of detergent added to the water appears to hasten disintegration. After saturation with gasoline (or kerosene), immersion into a boiling 50% hydrous solution of sodium carbonate (washing soda) and continued boiling has been suggested for some porous limestones as well as well indurated shales.

11. Heat the sample, then plunge it into cold water. (Some grains may be fragmented; oxidation and other effects on some minerals may be expected. Size analysis, and probably mineralogical analysis, of clays cannot be performed.)

12. Alternately freeze and thaw samples immersed in water. (This technique is effective, but it usually requires a long time and is useless with impervious rocks such as well-indurated mudstones. It may damage some crystal structures.)

13. Cover the crushed fragments with an equal amount of sodium acetate, add a few drops of water, and heat. The acetate will melt and saturate the sample; adding a crystal of acetate during cooling causes general crystallization and should disintegrate porous rocks.

14. Dry the sample and saturate it with either a concentrated solution of sodium hyposulfite or 14% sodium sulfate decahydrate (Glauber's salts). Crystallize the compound by drying.

15. Dry the sample and immerse it in a melt of sodium thiosulfate ("hypo"). Boil gently. Remove the sample and allow it to cool thoroughly. (Hypo crystallizes and disaggregates the sample but cements particles.) Recycle as necessary—grains will accumulate in the container. (If you use a beaker, don't allow hypo to crystallize in it—reheating will break the glass.) Sodium thiosulfate is very soluble in water, and repeated washing will remove it. (Clays may be retained on filter paper, but cation exchange is likely to have taken place.)

16. Saturate the sample with calcium bicarbonate, or boil it for a few moments in a sodium carbonate solution. Subsequent immersion in dilute hydro-

chloric acid causes effervescence, which may disintegrate the rock.

17. Using an iron mortar and pestle, pound the sample. (Remove fines periodically, and remember that iron fragments must be removed from the final sample either by using a magnet, which will also remove magnetite and ilmenite, or by dissolving them in hot nitric acid, which may affect other minerals.)

18. Crush the sample in a mechanical disc pulverizer. (This is a last resort; the sample will then be suitable only for such investigations as chemical analysis, trace element analysis within certain minerals, and limited mineralogical analysis. Fragments of the discs must be eliminated. Evaluate the results by comparing with portions of the original sample and thin sections).

REMOVAL OF CEMENTS BY CHEMICAL MEANS

Caution: Extreme care is needed when using some of the chemicals—always work under a fume hood and near a water supply for washing. Be familiar with the properties of the chemicals and reaction products and their potential effects on skin and respiration!

Be aware of the effects various chemicals may have on minerals (see Krumbein and Pettijohn, 1938, p. 355). After putting clays through many chemical treatments, size or mineralogical analysis is useless. If clays are to be retained, it may be necessary to flocculate them by adding NaCl so that decantation of the chemical reaction products may be done (an involved filtration procedure to clean the clays of electrolytes may be necessary as well).

A preliminary mechanical disaggregation is necessary to produce rock fragments 1 cm or smaller before most chemical treatments are applied.

Quantitative measurement of sample losses from chemical action can be achieved with proper experimental design. (Commonly it is sufficient to measure the amount of weight lost by weighing before and after treatment, but reaction precipitates must be removed.)

To Remove Carbonates

1. For the least effect on other minerals, Jackson (1958) recommends sodium acetate treatment for calcium carbonate and gypsum (ineffective with large crystals or concentrations). Brewer (1964) suggests the following procedure: place 100 g fine sample in a 3-l beaker, add 600 ml $2N$ sodium acetate solution (buffer for pH 5) and stir into suspension; allow to digest on water bath. Allow settling, decant the fluid, and wash with sodium acetate solution to remove excess calcium salts, then wash with distilled water and decant until clear. Repeat as necessary. (To avoid oxidizing organic matter, bubble SO_2 gas through the suspension overnight.)

2. When delicate components might be damaged by effervescence, use dilute acetic acid (approximately 15%) or monochloracetic acid (50%), which is faster but be sure to *avoid contact with your skin*. (With acetic acid, the sample must generally remain for days in the solution—until no reaction occurs when fresh acid is added.)

3. Use warm dilute HCl (10–15%) for dolomite, magnesite, and siderite (this may not be fully successful unless the sample is very fine). For the least harmful action on other minerals, use the smallest amount of HCl possible. Brewer (1964) suggests adding 100 ml $2N$ HCl to 1 l water, plus 5 ml $2N$ HCl for each percentage point of calcium carbonate above 2%. Expanding clays are markedly affected.

4. Boil the sample gently in 20% phosphoric acid for 5 min. This may not remove abundant calcite cement, but it may remove iron carbonates and does remove iron oxides and phosphates such as apatite. A finely crystalline calcium phosphate will precipitate (see Leith, 1950). Oxalic, formic, citric, and sulfuric acids have also been used, and any one may be particularly effective for a sample suite.

5. Organic complexing agents that form soluble complexes with alkaline earths may be used (for example, EDTA, or ethylene-diamine-tetra-acetic acid). A 10% EDTA solution is recommended, using 50 ml solution for each 0.5–1 g limestone. Boil until the carbonate is dissolved (that is, until a small suspended sample does not react to HCl). pH levels are critical for iron removal (low pH) and to keep clays from altering (high pH; raise by adding NaOH). Other calcium, strontium, and barium minerals (such as apatite, barite, and gypsum) are dissolved. (See Glover, 1961; Bodine and Fernald, 1973).

To Remove Sulfides and Sulfates

1. *Pyrite:* use warm (to boiling) 15% nitric acid (which also removes phosphates), oxalic acid, or hydrogen peroxide (15%).

2. *Pyrrhotite:* use dilute HCl.

3. *Calcium sulfates:* use warm concentrated HCl. Use dilute HCl if gypsum is fine (for drastic removal of gypsum, use a strong ammoniacal solution of ammonium sulfate).

4. *Barium sulfate:* use warm concentrated H_2SO_4 (which also removes chlorite and biotite and affects other minerals).

To Remove Organic Matter (as is necessary before size analysis of modern sediments)

1. Use hydrogen peroxide (ensure that the calcium carbonate content is negligible, or else excess calcium oxalates will form). Treat cold or warm, or boil in 10% solution, or treat cold in 30% H_2O_2 (see, for example, Jackson, 1958). Efficiency is increased by adding a few drops of lN HCl. Initially, just cover the sample with distilled water, then add small quantities of H_2O_2, stirring until any effervescence ceases. (Make sure the beaker is large enough that frothing will not cause overflow—for example, a 250 cm^3, high beaker for 50 g sediment.) A large watch glass cover may be necessary if frothing is excessive. If the solution heats excessively, cool the beaker in a water-filled container. Continue adding H_2O_2 until frothing ceases, then heat slowly to 60-70°C (H_2O_2 decomposes above 70°C). Observe for 10 min to ensure that any danger of a strong reaction has passed. Add H_2O_2 until no further reaction occurs. Elemental carbon and paraffin-like compounds are unaffected; coarse fibrous organic material will have to be removed by hand.

2. Use the sodium hypobromite method (see, for example, Brewer, 1964): for 100 g (of soil), add 400 ml fresh sodium hypobromite solution (mix equal volumes of bromine water and 40% NaOH solution). Stir and let the suspension stand 2 hr at room temperature with occasional agitation. Add another 400 ml of solution, stir, and let stand overnight (12 hr). Wash by decanting or filter.

3. For bituminous matter (such as cements), use carbon disulfide, ether, or chloroform, washed with carbon tetrachloride, acetone, or alcohol. (Bituminous matter fluxes with hot Canada balsam; if thin sections are to be prepared of bituminous sediment, special precautions must be taken—see Milner, 1962.) Solid asphalts and coal must be treated specially.

To Remove Iron Oxides (try ultrasonic treatment first)

1. To cause the least harm to other sediments (such as clays—even bentonite is reported unchanged): add 300 ml distilled water to approximately 100 g of sample in a beaker. Add 24 g solid sodium citrate + 2.8 g solid sodium bicarbonate (the pH should remain at 7.0-7.5). Heat to 75-80°C while stirring to ensure complete solution. Add 7g solid sodium hyposulfite (sodium dithionite, $Na_2S_2O_4$ + H_2O), stir constantly for 5 min and intermittently for 10 min. Wash by repeated sedimentation and decantation or filter with a Buchner funnel (Brewer, 1964; see also Mitchell and MacKenzie, 1954). Warming (with occasional shaking) for 20-30 min may work equally well after adding 20% sodium dithionite solution to a disperse suspension with a little water. The sample should turn gray. Wash it several times with 1% neutralizing NaCl solution in this case before washing with distilled water. (*Note:* Dithionite solutions turn brown and lose their effect rapidly—hence always use fresh solutions.)

2. Warm the sample, which must be finer than pea-size, in 1N (100 ml standard 35.4% HCl solution + 900 ml distilled H_2O), 6N (100 ml in 400 ml H_2O), or 50% HCl solution until colorless.

3. Warm with 15-20% HCl + 10% stannous chloride (or 5 g stannous chloride in 10 ml HCl) until colorless (Drosdoff and Truog, 1935).

4. Treat in oxalic acid and sodium sulfide (for the procedure, see Truog et al., 1936). This also aids in removing other cementing materials. Alternatively, shake (30 min) or boil (10 min) in a solution of 31.5 g oxalic acid × 62.1 g ammonium oxalate in 2.5 l H_2O.

5. Boil gently for 5 min in 20% phosphoric acid; this will also remove phosphates and carbonate cement.

6. Place the sample (approximately 20 g) in a 500 ml beaker. Add 300 cm^3 distilled water then 15 g powdered oxalic acid and a cylinder of sheet aluminum (with a diameter slightly smaller than that of the beaker, extending approximately 2 cm above the fluid from the base). Boil gently for 20 min. Remove the cylinder and decant the liquid. Wash the sample and decant the wash water until it becomes clear (see Leith, 1950). This procedure has a very slight effect on most minerals, but it probably affects clays. A yellow ferrous oxalate should precipitate on the aluminum cylinder.

7. Remove any carbonates. Place 2 g sample in a 150 ml beaker, add 40 ml potassium oxalate solution (103.7 g/l), and heat to 80°C. Add 10 ml oxalic acid solution (95 g/l), stir, and heat to 90°C. Add approximately 0.2 g of magnesium ribbon (wrapped around a rod) and stir for 3-5 min at 90-95°C. Add 5 ml oxalic acid solution, and continue heating for 3 min or until the material turns white or gray. Wash by repeated sedimentation and decantation, or by filtering. For first washing, use N/10 HCl (Brewer, 1964).

To Remove Silica Cement

1. Hot strong alkali solutions (KOH or NaOH) may dissolve opal, chalcedony, chert, or authigenic quartz overgrowths.

2. Hydrofluoric acid may be used to extract some heavy minerals. (Add crushed material to cold concentrated HF in a platinum or lead crucible. Heat is generated.) Be careful!

DISPERSION OF CLAYS

Although clays do not normally settle in natural environments as individual particles, complete dispersion is necessary for textural analysis in the laboratory to avoid inconsistent results and to yield comparable data. Even then, the "true size" of the clays is not measured because the analytical techniques do not differentiate between variations in size, shape, and density (density depends on mineralogy and the extent of cation adsorption).

For size analysis of clays, strict standardization in pretreatment as well as measurement technique is necessary for all samples to be compared—be wary of comparing results of different analysts or laboratories and clearly report your procedure.

The most effective dispersion agents (peptizers) for clays are strong bases, weak acids, and large acidic anions. Any dispersant will cause flocculation (that is, act as a coagulant) if its concentration is sufficiently high or if it is allowed to act on the clays long enough. In fact, dispersants cause rapid dispersion at first and slow, continuous dispersion (or coagulation) later—total dispersion may never really be achieved during an analysis. The presence of a significant quantity of natural electrolytes in the rock sample may invalidate the use of some dispersants, and if the quantity is large, you may

have to subject the sample to a preliminary washing and filtration procedure. In addition, the various minerals (or size ranges) present may dictate the type and concentration of peptizer you use, as well as the time necessary for dispersion—trial-and-error experimentation for "best" results may be necessary. To test for flocculation, let the suspension stand for up to 12 hr, tilt the beaker of suspension approximately 20°: if settled sediments or stratified layers flow to maintain a horizontal level, flocculation has occurred. You may also examine a drop of the suspension under a microscope—particles should be separate and the finest should show Brownian movement. *For maximum dispersion, it is probably best to subject the suspension to 15 min in an ultrasonic device as a final step.*

Note: In most cases, correction factors must be applied to the results of the analysis to allow for the weight of the added peptizer.

The following are some common peptizers (with their concentrations in distilled water, see also Anderson, 1963; Tschillingarian, 1952; Krumbein and Pettijohn, 1938):

ammonium (or sodium) hydroxide (a few drops of concentrated $NH_4(OH)$ per liter); this peptizer will not work if calcium or magnesium carbonates are present

sodium carbonate (0.53-2.12 g/l, try 1.06 g/l first)

sodium oxalate (0.67 g/l, or 0.34 g/l)

sodium hexametaphosphate (0.25-0.6 g/l), known by the trade name Calgon.

In general, avoid oven-drying of clays if you plan to attempt size analysis. (It should, however, be all right to dry them at 60-65°C in a drying cabinet.) Also try to avoid boiling as an aid to dispersion, but doing so may be necessary in recalcitrant cases. For upper Cenozoic samples, it appears best to disperse the sample in the state in which it was collected (that is, do not dry it; see Prokopovich and Nishi, 1967), but washing to remove soluble salts may still be required.

In addition to disaggregation methods mentioned previously, the following techniques have been applied to clay-rich samples:

Boil the sample in a caustic soda (NaOH) solution (20%) for several hours.

Place caustic soda pellets on the sample, and allow hydration from atmospheric moisture to take place.

Add boiling water to dry washing soda (Na_2CO_3), covering the sample.

Boil the sample in a $0.01N$ ammonia solution or a $0.01N$ sodium-pyrophosphate solution.

Soak the sample in a hydrogen peroxide solution. Wick (1947, in Mueller, 1967) suggests the following procedure: Reduce the dry sample to fragments 3-6 mm in size; pour 15% H_2O_2 solution (dilute 30% commercial solution with water in 1:1 ratio) over the fragments in a large beaker under a hood (just immerse the fragments). Boiling will occur within a few minutes and water vapor plus oxygen will be generated by fluid dissociation. (If boiling has not begun within 10 min, heat the sample and add a few milliliters of KOH). Repeat the process for strongly indurated rocks. Wet-sieve and recycle the remaining aggregates.

WASHING AFTER CHEMICAL TREATMENTS

To cleanse sediments after chemical methods of disaggregation, and to cleanse loose modern sediments (such as soil or marine samples that have electrolytes present), use a sedimentation-plus-decantation method (if particles are not too fine), or wash and filter using an apparatus as shown in Fig. 50.

Apply the vacuum and wet the filter paper to achieve a firm seal. Place the sample on filter paper and apply the vacuum while washing with distilled water and stirring with a glass rod. Dry the sample at 40-60°C on the filter paper before proceeding to further analysis.

Note: The rubber hose should flatten when adequate vacuum has built up, at which time you should apply the clamp and turn off the pump. When the vacuum dissipates, open the clamp and turn the pump on again.

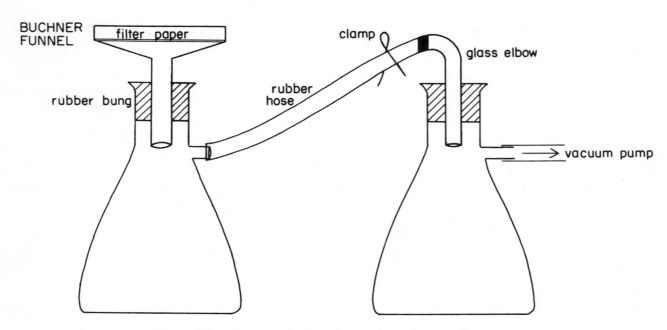

Figure 50. Suggested setup for washing dispersed samples.

Grain-Size Analysis

All grain-size analyses involve indirect methods of measuring size and are biased by the variable of shape (for example, the length of intermediate axis determines which grains pass through a sieve mesh), and in settling-tube analyses, by particle density as well. Thus care must be taken to compare only samples that do not differ markedly in grain shape and composition, and to ensure that disaggregation is complete.

Depending on the characteristics of the sample suite or the statistical parameters to be calculated, it may be appropriate to ignore "tails" of the distribution when they total less than *ca.* 5 wt % (using the Folk and Ward parameters for example). Hence sieve analysis of a sample comprising 95 wt % mud, or pipette analysis of a sample comprising 95 wt % sand may not be warranted.

REPRESENTATIVE SUBSAMPLING

Samples must be large enough to give statistically meaningful results, hence must be "large" relative to the largest particle size present. Fig. 51 is a guide to the minimum sample required for sieve and/or pipette analysis. Conversely, sieves should not be overloaded with sediment to avoid damaging the mesh; the table in Fig. 51 shows maximum values. Nor should 1-liter settling cylinders be loaded with much more than 10–25 g for pipette analysis, or 30–40 g for hydrometer analysis, to avoid interference effects. The inset in Figure 51 shows total subsample weights for sand/mud mixtures that will provide suitable quantities of mud for pipette analysis. Careful splitting of samples to appropriate subsample weights becomes more critical with increasing grain size range; that is, a large subsample may be necessary for gravel analysis, a smaller one for sand analysis, and an even smaller one for mud. Keep a record of the weight of each subsample and the initial total sample!

Grain-size segregation commonly occurs during transport of sample bags—ensure that your subsample is representative of the whole by thorough mixing prior to subsampling. Dry the sediment, mix it thoroughly, then pour it evenly into the hopper of a sample splitter, repeating until a sufficiently small subsample is obtained. Equally satisfactory, and applicable with wet sediments, is successive quartering of cones on a piece of glazed or waxed paper: flatten the cone, quarter with a spatula, reject two opposite corners, mix the other two quarters and repeat as necessary. For most homogeneous sandy or muddy sediments, a tablespoon blindly dipped into the mixture is an adequate subsampling technique, as long as differential segregation has not occurred.

PROCEDURE FOR SIEVE SIZE ANALYSIS OF A SAND

1. Select a representative subsample and label a data sheet for it (see Fig. 52).

2. (a) For samples with less than about 10% mud and when analysis of the mud is not necessary, dry the subsample at no more than 65°C (to avoid baking clays). Leave it to cool and equilibrate with the atmosphere for at least 1 hr before weighing it to 0.001 g. Thoroughly disaggregate the sample—for most loose sands, a rubber bung on a piece of glazed paper is adequate; otherwise, see "Pretreatment of Samples."

(b) For samples with a mud fraction to be analyzed, wet-sieving is necessary. If the mud fraction is to be analyzed by pipette, the total dry sample weight is not required (although you may wish to have it to check experimental error). For the wet-sieving procedure, see the following section "Pipette Analysis of Mud," steps 3 and 4. After wet-sieving, dry the coarse fraction and weigh it to 0.001 g; the sand is now ready for sieving.

3. Select a nest of sieves to cover the grain size range of the sample. The mesh in the coarsest sieve should equal the size of the coarsest grains. If the sample has been wet-sieved, the finest sieve should be 4∅; otherwise you may use sieves as fine as 4.75∅. For detailed work and where you suspect polymodal distributions, use ¼∅ intervals.

4. Clean the sieves before using them: invert each sieve and tap it *evenly* onto a flat surface or, using your hand, rap the side *diagonally* to the mesh to knock out any loose grains. Then brush the screen, again *diagonally* to the mesh, with a soft sieve brush. If any grains are trapped in the mesh, do not attempt to force them out—leave them there. Stack the sieves in the correct order, with the pan at the bottom. If you must make two nests, use the coarser set first, then transfer the contents of the pan to the finer stack (with another pan under it!).

5. Pour the sample into the top sieve and add the cover. Secure the sieve nest firmly in the sieve shaker. Shake for a standardized time—15 min is normal. (How many grains pass a mesh in a given

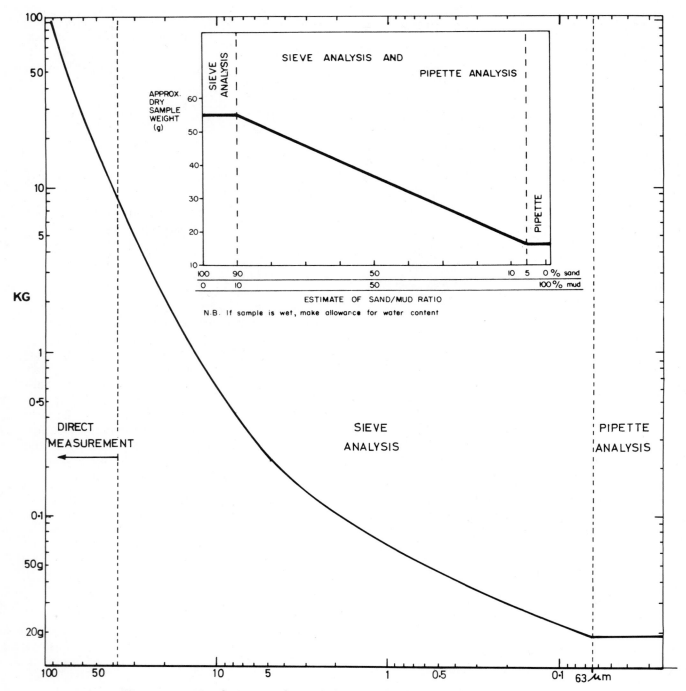

Figure 51. Guide to sample sizes for standard grain size analytical techniques. *(Diagram by G. Coates)*

DATA SHEET FOR SIEVE ANALYSIS

Sample no. _____ Analysed by _____ Treatment _____ Date _____

Particulars _____

Weights : dry sample _____ sand _____ mud _____ sand & mud _____

sieve diam. Ø	exact diam. Ø	weight beaker	weight beaker and sample	weight sample	% aggs.	corr- ected weight	cumul- ative weight	cumul- ative %	% shell	notes
-5.00										
-4.00										
-3.00										
-2.50										
-2.25										
-2.00										
-1.75										
-1.50										
-1.25										
-1.00										
-0.75										
-0.50										
-0.25										
0.00										
+0.25										
+0.50										
+0.75										
+1.00										
+1.25										
+1.50										
+1.75										
+2.00										
+2.25										
+2.50										
+2.75										
+3.00										
+3.25										
+3.50										
+3.75										
+4.00										
+4.25										
+4.50										
+4.75										
pan										
Total										

$$\text{Percent error} = \frac{(\text{original sample weight} - \text{total weight retained}) \times 100}{\text{original weight}}$$

Figure 52. Data sheet for sieve analysis.

time is a probability function, and standardized shaking time is essential.)

6. After shaking, invert and clean each sieve as in step 4; retain each fraction on a large sheet of paper (preferably glazed), and transfer each to a labeled, preweighed beaker or envelope. If the sample has previously been wet-sieved and mud analysis is to follow, material passing the 4ϕ sieve (pan fraction) should be added to the mud fraction.

7. Weigh the beakers and sediment to 0.001 g and retain each fraction in a labeled envelope for future use.

8. Check each fraction for grain aggregates and obvious properties under a binocular microscope (for instance, look for proportions of components such as heavy minerals or rock fragments—there may be significant differences between fractions). If aggregates are common, either disaggregate and resieve, or carefully estimate the percentage of aggregates in each fraction and subtract this percentage of the weight of the fraction from both the weight of the fraction and the total weight of the subsample.

9. Compute the weight percentage of each fraction then compute cumulative percentages. The weight percent of each sand fraction is:

$$100 \times \frac{\text{wt. of sand on sieve}}{\text{total sample wt. (sand and mud)}}$$

Add these percentages incrementally to obtain cumulative weight percentages.

10. Plot the data in a cumulative curve on graph paper. Examples of graph papers are given in Figures 53-56, on pages **89-92**. Consistent "kicks" at the same size grade in graphs of different samples may indicate a defective sieve.

GRAIN SIZE ANALYSIS OF MUD

A variety of techniques may be used to analyze the size of silt and clay fractions. Pipette analysis, which has been widely used for a number of years, has a sound theoretical basis. Precision (reproducibility) is high—better than 0.1ϕ unit—but the method is time-consuming. Size analysis by electronic instruments is considerably easier and faster, but precision of the particular instrument must be tested. As long as precision is high, the efficiency of the electronic methods makes them ideal for internal studies. Size analysis by hydrometer gives results that compare well with those of pipette analysis, and reproducibility is almost as good as that of pipette analysis (see, for example, Fig. 57 and Kaddah, 1974). The method is considerably easier and quicker to carry out than pipette analysis.

It should be realized that any method of size analysis is an indirect method of measurement. Different indirect methods will yield different results, which cannot be compared directly with one another. For example, experiments with one electronic size analyzer, a hydrophotometer, give results that are consistently coarser than those of pipette analysis. Another electronic instrument gives results that are finer over much of the silt fraction (see Fig. 57 on page **95**).

Regardless of the technique you use, you should carry out a size analysis only when disaggregation and dispersion have not substantially modified the physical or chemical properties of clays. Swelling clays present a special problem, and it is doubtful whether useful data can be obtained from sediments that have been diagenetically modified. Also remember that laboratory dispersion techniques produce a "size" distribution that may not be realistic in terms of natural sediments (clays in nature travel as aggregates or flocs).

Pipette Analysis of Mud

General

Pipette analysis is based on calculations of the settling velocities for particles of different "size" (Stokes' Law); the generally invalid assumption is made that all particles are of the same shape and density (s.g. 2.65). Subsamples of a specific volume are extracted from a suspension of mud at specified times and depths; the weight of each dried subsample is representative of the proportion of the total mud fraction remaining in suspension above that specified depth at that specified time. Thus each subsample measures the proportion of total mud that is *finer than* the "size" that will have settled to the specified depth in the specified time. This concept is illustrated in Fig. 58 on page **96**, which plots grain size against a log time scale: at a particular time, you can read off the maximum grain size that will have settled to the specified (10- or 20-cm) depth. For example, after 2 hr, the maximum grain size 10 cm below the surface is 8ϕ and a withdrawal at this depth will contain sediment finer than 8ϕ.

Sample no. _____ Analyst _____ Date _____

Description of sample _____

Modal grade: _____ to _____ mm. Secondary modes: _____ to _____ mm.

_____ to _____ mm.

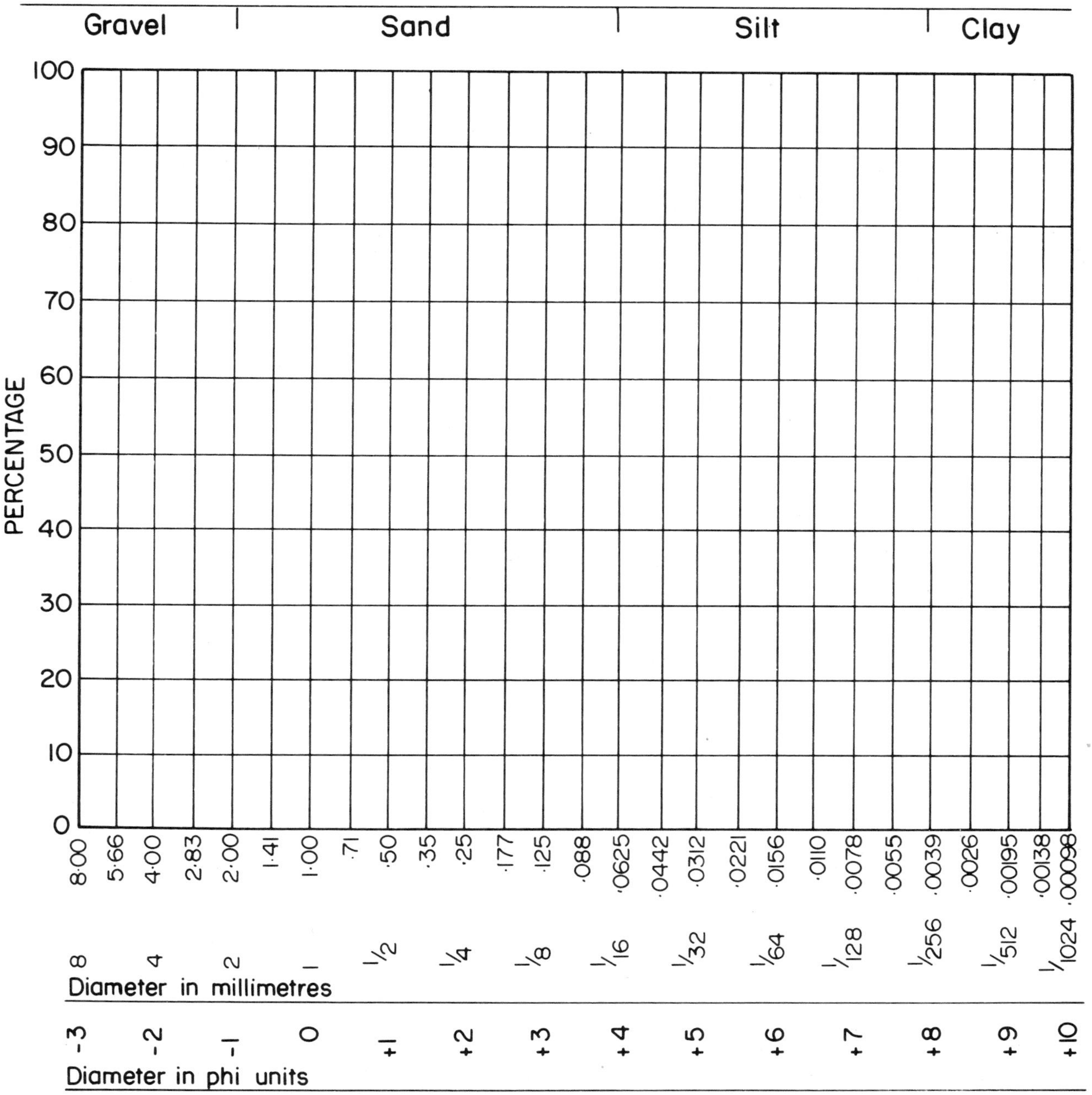

Figure 53. Graph paper for plotting grain-size data as histogram.

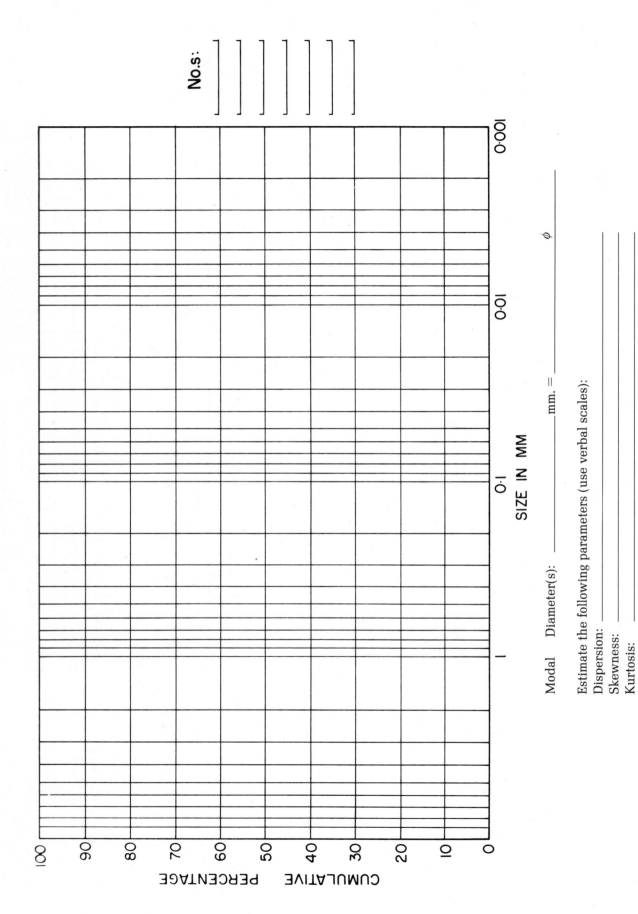

Figure 54. Semi-log paper for plotting grain-size distributions.

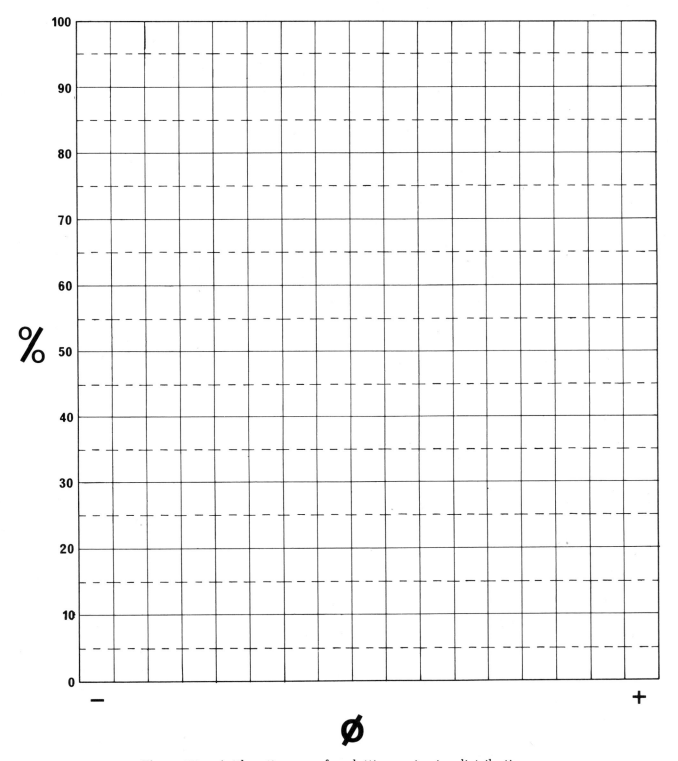

Figure 55. Arithmetic paper for plotting grain-size distributions.

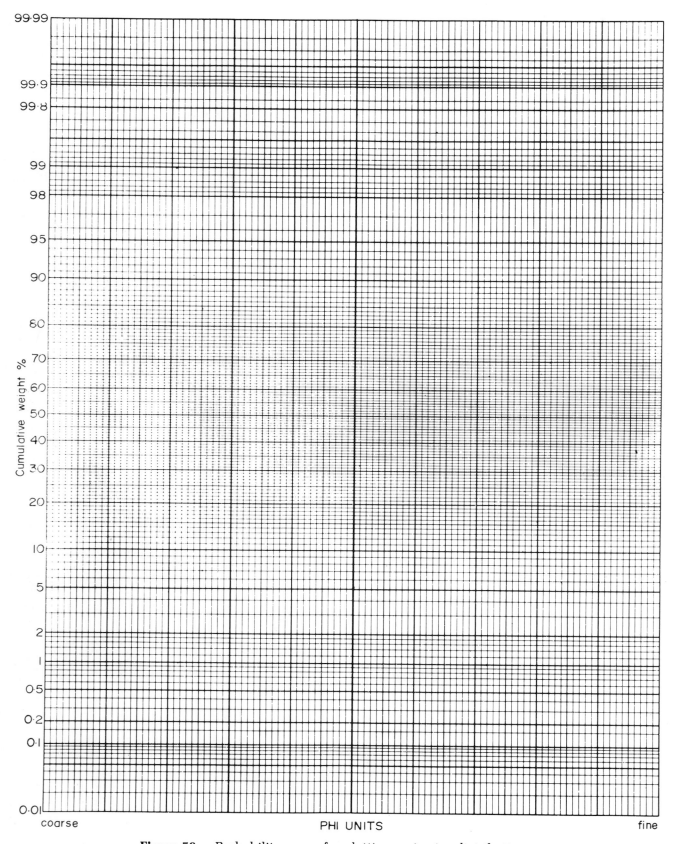

Figure 56. Probability paper for plotting grain-size distributions.

Table 9 (pages **97-98**) presents a time schedule for pipette analyses run of up to 21 samples in a day. Study the schedule, decide how many samples are to be run in one day, and plan your day so you will be free when withdrawals need to be made.

Temperature affects the viscosity of water and therefore settling velocities. A correction for the temperature effect can be made indirectly by altering the depth of withdrawal (Table 10, page **99**). Note the temperature of the water in the settling columns before and during any pipette analysis; change your sampling depth as necessary.

Equipment

For each sample, you will need the following:

1 data sheet
1 beaker or basin for disaggregation
one 1-l measuring cylinder
8 or 9 50-ml beakers
1 watch glass for cylinder

In addition, you should have the following general materials on hand:

4∅ wet sieve (do not use 4∅ dry sieve)
large evaporating basin
wash bottle with distilled water
large funnel (about 20 cm in diameter)
solution of dispersant (e.g., 50g/l "Calgon")
brass stirring rod
thermometer (0-100°C)
20-ml pipette with depth graduations (with insoluble ink, mark depths of 10 cm and 20 cm from lower tip of pipette; a few additional marks at 0.5 cm intervals around those depths may also prove useful—see Table 10)
rubber pipette filler
timepiece with hours, minutes and seconds
pipette analysis schedule and temperature/withdrawal depth sheet (Tables 9 and 10)
oven for drying mud fractions (60-65°C)

Preparation

You will need to allow up to one full day for preparing the samples, depending on how many you plan to analyze. The following procedure describes the preparation of one sample. Label a separate data sheet for each sample (see Fig. 59, page **100**).

1. Obtain a representative subsample that will yield no more than 15-20 g of mud.

2. Fully disaggregate the subsample (see "Pretreatment of Samples"). It is often adequate to cover the sample with a little distilled water in a beaker and to use your fingers in a rubber glove to break up the sample fully. (Adding a solution of dispersant at this stage instead of later, in step 6, may assist disaggregation.) Rinse mud off your glove back into the beaker.

3. Wet-sieve the sample with a 4 ∅ (0.063-mm) wet sieve. (Do not use a 4∅ sieve from a set of dry sieves.) Place the sieve over a large evaporating basin and wash *all* its fines into the sieve using as little distilled water as possible—you must end up with no more than 900 ml of water and mud! (After you have about 600 ml, let the silt settle out, then use the partly clear water for further wet-sieving; wash finally with clean water.)

4. Transfer all the sand fraction retained on the sieve to an evaporating basin or a beaker, using the wash bottle. Dry the sand fraction, leave it to cool for 1 hr, and weigh it to 0.001 g. If there is a significant amount of sand, you will need to sieve it *before* carrying out the pipette analysis, because more mud fraction may appear after dry sieving.

5. Transfer all the mud collected in the basin to a 1-l measuring cylinder via a large funnel (label each cylinder).

6. Add dispersant to the column if you have not previously used a solution with dispersant in your wash bottle or for disaggregation. Between about 0.5 and 1 g of sodium hexametaphospate ("Calgon") is normally sufficient to prevent flocculation of clays. It is essential to know the exact amount of dispersant in each column for later calculations. A useful procedure is to make up a stock solution of 50 g/l; 20 ml of this solution (use a pipette) added to the column adds exactly 1 g of dispersant.

7. Top the column up to 1000 ml with distilled water. Thoroughly stir the column using a brass stirring rod.

8. Label, and weigh to 0.001 g, 8 (or 9) 50-ml beakers (one for each withdrawal on your pipette data sheet). Arrange the beakers in front of the column.

9. Cover the column with a watchglass and let it stand overnight to check for flocculation before running the pipette analysis. At this point is is also useful to fill a beaker with tap water and insert a thermometer (preparatory to the next step).

Analysis

You will need to start the pipette analysis early in the morning, since the time between first and last withdrawls is at least 8 hours.

Before you begin, check that no columns have flocculated. Flocculation can be recognized by a curdling and rapid settling of clumps of particles, or by the presence of a thick, soupy layer on the bottom of the cylinder which passes abruptly into relatively clear water above. If flocculation is evident, try adding more dispersant solution or make up a new suspension with a smaller amount of sample. Using a mechanical stirrer for 5 minutes may assist dispersion.

10. Take the temperature of the water in the beaker of tap water and look up the corrected depths in Table 10. Note these depths on the pipette schedule, and monitor any temperature changes during the analysis.

11. Select a 20-ml pipette (one that empties quickly) with depth graduations as described in the equipment list. Connect a rubber pipette filler to the pipette and check that the suction works efficiently. Have a large beaker of distilled water ready on the bench for rinsing.

12. Start the timepiece 1 min before the initial withdrawal (if you are using an electronic timepiece, set it at 11:59 P.M.). Immediately begin stirring column 1, using a brass stirrer. Start with short, quick strokes at the bottom and stir up all the settled mud, then work up the column with long, vigorous strokes, being careful not to mix air in with the suspension. Precisely at time zero (12:00:00 on the electronic timepiece), withdraw the stirrer. Lower the pipette to 20 cm. At *exactly* 20 sec, extract a *20-ml* sample. Empty it into the respective 50-ml beaker and then rinse the pipette into the same beaker after sucking up 20 ml distilled water.

This first withdrawal is particularly critical since it represents everything finer than $4\emptyset$ (that is, total mud).

13. The next withdrawal is for the fraction finer than $4.5\emptyset$. At exactly 2 min, withdraw 20 mls, empty it into the next beaker, and rinse as before.

Repeat the procedure for all subsequent withdrawals. Efficiency is essential, particularly where multiple samples are to be analyzed. Initially, a withdrawal must be made and the next column stirred within 1 min. Withdrawal and rinsing need to be completed in 30 sec, leaving 30 sec for stirring the next column. To ensure thorough stirring of the column, make a preliminary stir in an earlier spare moment.

If you make a withdrawal at the wrong depth or wrong time, make a note of the error. You can then use Fig. 58 to find the grain size represented.

When there are long periods between withdrawals, cover each column with a watch glass. Any external source of vibration must be eliminated during the analysis.

14. When all withdrawals are completed, put the beakers onto trays and dry them in an oven. If further analysis of the clays is to follow, do *not* heat them above 65°C. It may take up to 48 hr to evaporate all the water.

15. When they are dry, remove the beakers from the oven and leave them to equilibrate with the atmosphere for at least 1½ hr. Weigh them to 0.001 g, recording the weights on data sheets.

16. Calculate cumulative weight percentages:
(a) Subtract beaker weights from beaker + sediment weights to get sediment weights.
(b) Multiply the weight of sediment from the $4\emptyset$ sample by 50 and subtract the weight of dispersant in the column. This gives the total weight of mud (F). For example:

$$[0.405 \text{ g } (4\emptyset \text{ sediment weight}) \times 50] - 1 \text{ g}$$
(wt. of Calgon in the procedure suggested) =
$$19.25 \text{ g (weight of mud, } F)$$

This value, added to the weight of the sand fraction (S) determined from step 4, provides the total sample weight. (To test for experimental error, total sample dry weight must be measured initially, but even low-temperature drying may cause problems in subsequent dispersion of the clay fraction.)
(c) Add the sand percentages cumulatively to obtain their cumulative percentages (step 9 of sieve analysis).
(d) Remember that each pipette sample represents material in the column finer than a certain grain size. To obtain cumulative percentages for mud intervals, multiply each mud weight by 50, subtract the weight of dispersant, divide by the *total* sample weight, and subtract from 100:

$$\text{cum.\% (mud range)} = 100 - \big([(50 \times \text{pipette sample wt.}) - 1 \text{ (assuming 1 g/l dispersant)}]/S + F\big)$$

A computer program can be used to process the raw data (Slatt and Press, 1976; Coates and Hulse, 1981).

17. Plot results on graph paper as required (see Figs. 53–56) or process by Method of Moments.

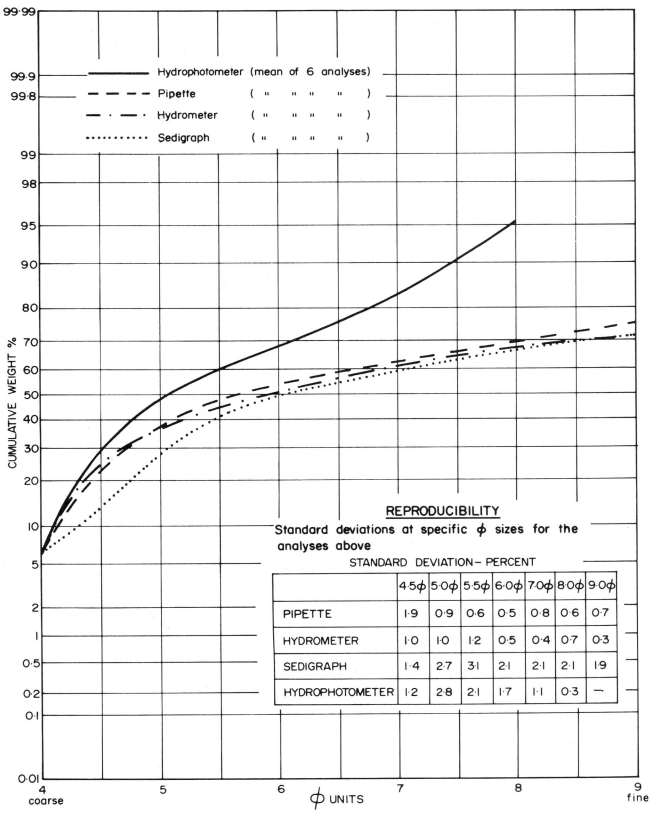

Figure 57. Grain-size distribution of a loess sample determined by four methods. *(Analysis and diagram by G. Coates)*

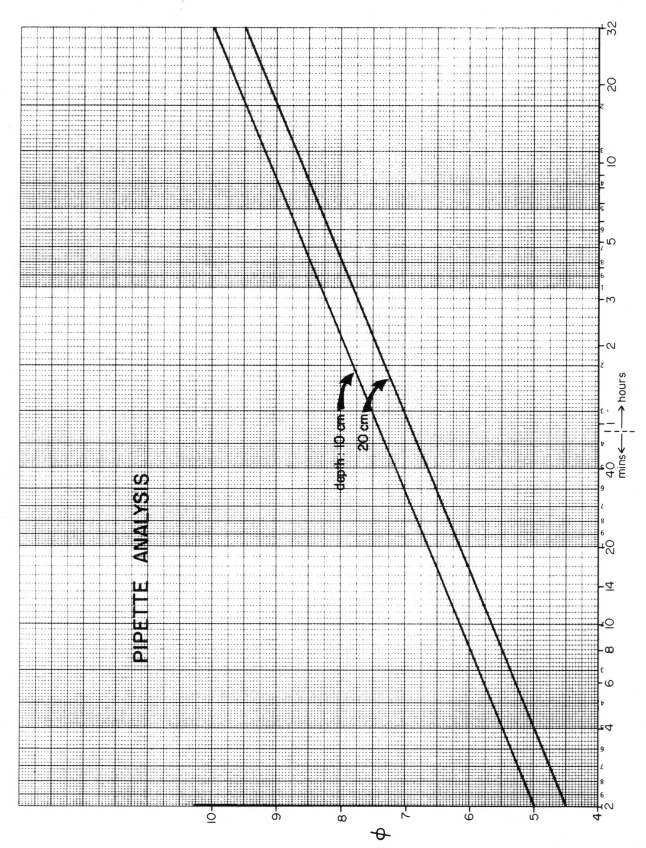

Figure 58. Plot of ϕ diameter versus time for pipette withdrawal at depths of 10 and 20 cm.

96

Table 9. Schedule for Carrying Out up to 21 Pipette Analyses in One Day, 8:00 A.M. to 8:14 P.M. (Lunch: 45 minutes; Dinner: 1 hour)

Time (hr:min:sec)	Sample No.	Approx. Depth (cm)	Size (ϕ)	Time (hr:min:sec)	Sample No.	Approx. Depth (cm)	Size (ϕ)
0:00	1	—	Start	53	9	—	Start
0:20	1	20	4.0	53:20	9	20	4.0
2	1	20	4.5	54	5	10	7.0
4	1	20	5.0	55	9	20	4.5
5	2	—	Start	56	8	20	5.5
5:20	2	20	4.0	57	9	20	5.0
7	2	20	4.5	58	7	20	6.0
8	1	20	5.5	59	6	10	7.0
9	2	20	5.0	1:01:00	9	20	5.5
10	3	—	Start	1:03	8	20	6.0
10:20	3	20	4.0	1:08	9	20	6.0
12	3	20	4.5	1:12	10	—	Start
13	2	20	5.5	1:12:20	10	20	4.0
14	3	20	5.0	1:13	7	10	7.0
15	1	20	6.0	1:14	10	20	4.5
18	3	20	5.5	1:16	10	20	5.0
19	4	—	Start	1:17	11	—	Start
19:20	4	20	4.0	1:17:20	11	20	4.0
20	2	20	6.0	1:18	8	10	7.0
21	4	20	4.5	1:19	11	20	4.5
23	4	20	5.0	1:20	10	20	5.5
24	5	—	Start	1:21	11	20	5.0
24:20	5	20	4.0	1:22	12	—	Start
25	3	20	6.0	1:22:20	12	20	4.0
26	5	20	4.5	1:23	9	10	7.0
27	4	20	5.5	1:24	12	20	4.5
28	5	20	5.0	1:25	11	20	5.5
29	6	—	Start	1:26	12	20	5.0
29:20	6	20	4.0	1:27	10	20	6.0
30	1	10	7.0	1:30	12	20	5.5
31	6	20	4.5	1:32	11	20	6.0
32	5	20	5.5	1:37	12	20	6.0
33	6	20	5.0	1:42	10	10	7.0
34	4	20	6.0	1:46	13	—	Start
35	2	10	7.0	1:46:20	13	20	4.0
37	6	20	5.5	1:47	11	10	7.0
39	5	20	6.0	1:48	13	20	4.5
40	3	10	7.0	1:50	13	20	5.0
43	7	—	Start	1:51	14	—	Start
43:20	7	20	4.0	1:51:20	14	20	4.0
44	6	20	6.0	1:52	12	10	7.0
45	7	20	4.5	1:53	14	20	4.5
47	7	20	5.0	1:54	13	20	5.5
48	8	—	Start	1:55	14	20	5.0
48:20	8	20	4.0	1:56	15	—	Start
49	4	10	7.0	1:56:20	15	20	4.0
50	8	20	4.5	1:58	15	20	4.5
51	7	20	5.5	1:59	14	20	5.5
52	8	20	5.0	2:00	15	20	5.0

Table 9. *(Continued)*

Time (hr:min:sec)	Sample No.	Approx. Depth (cm)	Size (ϕ)	Time (hr:min:sec)	Sample No.	Approx. Depth (cm)	Size (ϕ)
2:00*	1	10	8.0	4:12	19	20	5.5
2:01	13	20	6.0	4:13	20	20	5.0
2:04	15	20	5.5	4:14	21	—	Start
2:05	2	10	8.0	4:14:20	21	20	4.0
2:06	14	20	6.0	4:16	21	20	4.5
2:10	3	10	8.0	4:17	20	20	5.5
2:11	15	20	6.0	4:18	21	20	5.0
2:16	13	10	7.0	4:19	19	20	6.0
2:19	4	10	8.0	4:22	21	20	5.5
2:21	14	10	7.0	4:24	20	20	6.0
2:24	5	10	8.0	4:25	16	10	8.0
2:25	16	—	Start	4:29	21	20	6.0
2:25:20	16	20	4.0	4:34	19	10	7.0
2:26	15	10	7.0	4:39	20	10	7.0
2:27	16	20	4.5	4:44	21	10	7.0
2:29	16	20	5.0	LUNCH BREAK			
2:29*	6	10	8.0	5:25	17	10	8.0
2:33	16	20	5.5	5:30	18	10	8.0
2:40	16	20	6.0	6:04	19	10	8.0
2:43	7	10	8.0	6:09	20	10	8.0
2:48	8	10	8.0	6:14	21	10	8.0
2:53	9	10	8.0	8:00	1	10	9.0
2:55	16	10	7.0	8:05	2	10	9.0
3:12	10	10	8.0	8:10	3	10	9.0
3:17	11	10	8.0	8:19	4	10	9.0
3:22	12	10	8.0	8:24	5	10	9.0
3:25	17	—	Start	8:29	6	10	9.0
3:25:20	17	20	4.0	8:43	7	10	9.0
3:27	17	20	4.5	8:48	8	10	9.0
3:29	17	20	5.0	8:53	9	10	9.0
3:30	18	—	Start	9:12	10	10	9.0
3:30:20	18	20	4.0	9:17	11	10	9.0
3:32	18	20	4.5	9:22	12	10	9.0
3:33	17	20	5.5	9:46	13	10	9.0
3:34	18	20	5.0	9:51	14	10	9.0
3:38	18	20	5.5	9:56	15	10	9.0
3:40	17	20	6.0	10:25	16	10	9.0
3:45	18	20	6.0	DINNER			
3:46	13	10	8.0	11:25	17	10	9.0
3:51	14	10	8.0	11:30	18	10	9.0
3:55	17	10	7.0	12:04	19	10	9.0
3:56	15	10	8.0	12:09	20	10	9.0
4:00	18	10	7.0	12:14	21	10	9.0
4:04	19	—	Start				
4:04:20	19	20	4.0				
4:06	19	20	4.5				
4:08	19	20	5.0				
4:09	20	—	Start				
4:09:20	20	20	4.0				
4:11	20	20	4.5				

Note: For 32-hour withdrawals (10 ϕ measures), add 24 hours to the times for the 9 ϕ withdrawals.

*Coincident withdrawal times: the 2-hour withdrawals will not be harmed by about 30 seconds delay.

Table 10. Depth (cm) of Pipette Insertion for Given Temperature
(Based on Particles with S.G. = 2.65) from Folk's (1974) formula,
after Stokes Law*

Temp. °C	Total Suspension 20 sec	4.5ϕ 44μ 2 min	5ϕ 31μ 4 min	5.5ϕ 22μ 8 min	6ϕ 16μ 15 min	7ϕ 8μ 30 min	8ϕ 4μ 2 hr	9ϕ 2μ 8 hr	10ϕ 1μ 32 hr
14.0	20.0	17.8	17.7	17.8	17.7	8.8	8.8	8.8	8.8
14.5	20.0	18.1	17.9	18.1	17.9	9.0	9.0	9.0	9.0
15.0	20.0	18.3	18.2	18.3	18.1	9.1	9.1	9.1	9.1
15.5	20.0	18.4	18.4	18.5	18.4	9.2	9.2	9.2	9.2
16.0	20.0	18.8	18.6	18.8	18.6	9.3	9.3	9.3	9.3
16.5	20.0	19.1	18.9	19.1	18.9	9.4	9.4	9.4	9.4
17.0	20.0	19.3	19.1	19.3	19.1	9.6	9.6	9.6	9.6
17.5	20.0	19.5	19.4	19.5	19.4	9.7	9.7	9.7	9.7
18.0	20.0	19.8	19.7	19.8	19.6	9.8	9.8	9.8	9.8
18.5	20.0	20.0	19.9	20.0	19.9	9.9	9.9	9.9	9.9
19.0	20.0	20.2	20.1	20.2	20.0	10.0	10.0	10.0	10.0
19.5	20.0	20.5	20.4	20.5	20.3	10.2	10.2	10.2	10.2
20.0	20.0	20.7	20.6	20.7	20.6	10.3	10.3	10.3	10.3
20.5	20.0	21.0	20.9	21.0	20.9	10.4	10.4	10.4	10.4
21.0	20.0	21.3	21.1	21.3	21.1	10.5	10.5	10.5	10.5
21.5	20.0	21.5	21.3	21.5	21.3	10.7	10.7	10.7	10.7
22.0	20.0	21.8	21.6	21.8	21.6	10.8	10.8	10.8	10.8
22.5	20.0	22.0	21.9	22.0	21.8	10.9	10.9	10.9	10.9
23.0	20.0	22.3	22.1	22.3	22.1	11.1	11.1	11.1	11.1
23.5	20.0	22.6	22.4	22.6	22.3	11.2	11.2	11.2	11.2
24.0	20.0	22.8	22.7	22.8	22.6	11.3	11.3	11.3	11.3
24.5	20.0	23.1	22.9	23.1	22.9	11.4	11.4	11.4	11.4
25.0	20.0	23.3	23.2	23.3	23.2	11.6	11.6	11.6	11.6
25.5	20.0	23.6	23.5	23.6	23.4	11.7	11.7	11.7	11.7
26.0	20.0	23.9	23.8	24.0	23.7	11.9	11.9	11.9	11.9
26.5	20.0	24.2	24.0	24.2	24.0	12.0	12.0	12.0	12.0
27.0	20.0	24.5	24.3	24.5	24.2	12.1	12.1	12.1	12.1
27.5	20.0	24.7	24.6	24.7	24.5	12.3	12.3	12.3	12.3

*Time (min) = Depth of withdrawal (cm)/(1500 $\times$ A $\times$ d²), where A is a constant for a given water viscosity and particle density and d is the particle diameter.

DATA SHEET FOR PIPETTE ANALYSIS

Peptizer _____

Amount peptizer for for 1000 ml. _____ gms.

Mud _____ gms

Sample No. _____

Dia-meter ø	Temp.	With-drawal depth (cm)	Time	Beaker No.	Wt. sample + beaker in gms.	Weight beaker in gms	Weight sample in gms	X50	Weight Fraction	Cumul. weight	Cumul. per-cent
+ 4.0			20 s.								
+ 4.5			2 m.								
+ 5.0			4 m.								
+ 5.5			8 m.								
+ 6.0			15 m.								
+ 7.0			30 m.								
+ 8.0			2 hr								
+ 9.0			8 hr								
+10.0			32 hr								

Figure 59. Data sheet for pipette analysis.

Hydrometer Analysis of Mud

General

Hydrometer analysis is based on the same principle as pipette analysis: that particles of different size will settle from suspension at predictable rates. The density of a sediment-water suspension is measured at intervals and the progressive decrease in density correlated with the size fractions that have settled past the measuring depth in the intervening times.

Whereas the actual measurements are quickly and easily made, care must be taken to calibrate each hydrometer initially and to correct for any extraneous variables that may affect fluid density (such as temperature fluctuations).

Equipment

Have the following materials on hand:

hydrometer (ASTM 152H)
mechanical shaker
thermometer (0–50°C
4∅ wet sieve
1000-ml graduated cylinder
20-ml pipette
stirring plunger
analytical balance (0.001g)
stopwatch
evaporating dish
500-ml shaking bottle
oven
wash bottle
250-ml measuring cylinder

Hydrometer Calibration

1. Immerse and steady the hydrometer in approximately 170 ml H_2O in a 250-ml cylinder; measure the water level before and after immersion.

2. Slowly withdraw the hydrometer until the water level is halfway between the two previous levels; clamp the hydrometer in position. Record the reading (R) on the hydrometer stem at the level of the cylinder top (one way is to lay a ruler across the cylinder for reference). See Fig. 60.

3. Measure the distance from the cylinder top to the water level (nearest 0.5 mm); this is distance (c) from R to the center of gravity of the hydrometer bulb (Fig. 60).

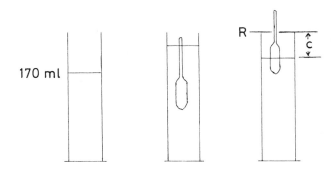

Figure 60. Hydrometer calibration—diagram illustrating steps 1, 2, and 3.

4. Remove the hydrometer; measure (to the nearest 0.5 mm) and record distances (y) from R to each graduation on the hydrometer stem (+y above R, −y below R). See Fig. 61.

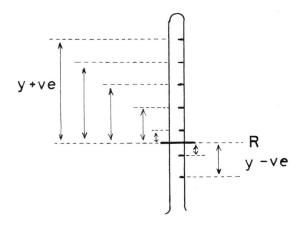

Figure 61. Hydrometer calibration, step 4.

5. Fill the 1-l cylinder to the 1000-ml mark; immerse the hydrometer and record the change in level (L) to the nearest 0–3 mm (Fig. 62).

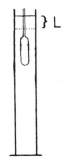

Figure 62. Hydrometer calibration, step 5.

6. Record the meniscus correction (Cm) as the difference between the top of the meniscus (R'h), determined from sighting on the cylinder wall, and the level water surface (Rh), read off the hydrometer stem. Hydrometers are normally read at the level of the liquid, but with opaque suspensions you can see only the top of the meniscus and the meniscus correction (Cm) must be added to the reading (R'h) to obtain the true reading (Rh). See Fig. 63.

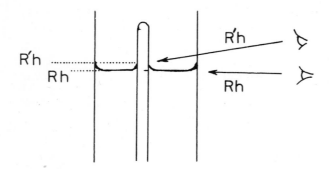

Figure 63.   Hydrometer calibration, step 6.

7. For each of the major graduations on the hydrometer stem, calculate:
(a) the effective depth (Hr) by $Hr = c + y - L/2$ (mm)
(b) the hydrometer reading at the top of the meniscus by $R'h = Rh - Cm$
Plot the calculated values of Hr and R'h on arithmetic graph paper; read the values of Hr from this graph for each reading of R'h you make with this hydrometer during subsequent analyses.

Specific Gravity Determination

To achieve greater accuracy or to compensate for samples of clearly differing composition, you can determine the specific gravity (S.G.) of the sample by the following method. For general purposes, an S.G. of 2.65 is assumed.

1. Dry 50–100 g of the sample in an oven; weigh it to 0.1 g.

2. Place the weighed sample into a bottle half-filled with water; insert a glass stopper and shake until dispersion is complete. Almost fill the bottle with water and allow the sample to settle for several minutes. Completely fill the bottle and insert the stopper. Dry the outside of the bottle and weigh it to 0.1 g.

3. Empty and wash bottle; fill with water, insert stopper, dry bottle and weigh to 0.1 g.

4. The solid density is calculated as follows:

$$\rho_s = \frac{\rho_w \times (\text{wt. of dry sample})}{(\text{wt. of bottle} + H_2O) + (\text{wt. of dry sample}) - (\text{wt. of bottle} + H_2O \text{ and sample})}$$

where ρw is the density of the water at the temperature of the test.

Preparation

Obtain a representative subsample of 30–40 g of mud and treat it as for pipette analysis (steps 2 and 3). Dry it in an oven at 65°C or less, let cool for 1 hr, and weigh to 0.001 g to obtain weight (W). Prepare settling cylinders as for pipette analysis (steps 5–7) and make sure the samples are dispersed (step 9).

Analysis

Table 11 shows a program for 21 analyses in one day. Readings are recorded on this sheet at the specified times. Record any departure from these times.

1. Fill a reference 1-l cylinder with water and dispersant solution as used for sample suspensions. Shortly before beginning the analysis, take a hydrometer reading in this cylinder at the top of the meniscus and record it as the composite correction (Cc), which, in the calculations, corrects for temperature, dispersing agent, and the meniscus. If the temperature varies by more than one degree during the analysis, remeasure Cc and note the time.

2. Stir each cylinder for 1 min. (as for pipette analysis, step 11) and take hydrometer readings at 1 min. and then at 2 min from the time you withdraw the stirrer (these times give approximately 4.5ϕ and 5.0ϕ size fractions).
To take readings: dry the hydrometer, then 20–30 sec before reading time, gently lower it into the suspension (allow 10-sec). Steady the hydrometer, then take a reading at the appropriate time. Remove the hydrometer gently (allow 10 sec), rinse it, and store it in a beaker of clean water.
Repeat the 1- and 2-min readings at least once more and enter the averaged readings on the data sheet (Fig. 64). (Repeated readings are necessary because the density of the suspension changes rapidly in the early stages of sedimentation.)

Table 11. Schedule and Data Sheet for Carrying out 21 Hydrometer Analyses in One Day

Time (hr. min)	Col. No.	R'h	Time (hr. min)	Col. No.	R'h	Time (hr. min)	Col. No.	R'h	Time (hr. min)	Col. No.	R'h
0.00	1	ST*	1.13	7		2.29**	16		4.19	4	
0.04	1		1.16	10		2.30	17	ST	4.19**	21	
0.05	2	ST	1.17	11	ST	2.33	16		4.24	5	
0.08	1		1.18	8		2.34	17		4.25	16	
0.09	2		1.19	4		2.35	18	ST	4.29	6	
0.10	3	ST	1.20	10		2.38	17		4.30	17	
0.13	2		1.21	11		2.39	18		4.35	18	
0.14	3		1.22	12	ST	2.40	16		4.43	7	
0.15	1		1.23	9		2.43	7		4.48	8	
0.18	3		1.24	5		2.43**	18		4.53	9	
0.19	4	ST	1.25	11		2.45	17		5.09	19	
0.20	2		1.26	12		2.46	13		5.12	10	
0.23	4		1.27	10		2.48	8		5.14	20	
0.24	5	ST	1.29	6		2.50	18		5.17	11	
0.25	3		1.30	12		2.51	14		5.19	21	
0.27	4		1.33	11		2.53	9		5.22	12	
0.28	5		1.38	12		2.55	16		5.46	13	
0.29	6	ST	1.42	10		2.56	15		5.51	14	
0.30	1		1.43	7		3.00	17		5.56	15	
0.32	5		1.46	13	ST	3.05	18		½ hr break		
0.33	6		1.47	11		3.09	19	ST	6.25	16	
0.34	4		1.48	8		3.12	10		6.30	17	
0.35	2		1.50	13		3.13	19		6.35	18	
0.37	6		1.51	14	ST	3.14	20	ST	½ hr break		
0.39	5		1.52	12		3.17	11		7.09	19	
0.40	3		1.53	9		3.17**	19		7.14	20	
0.43	7	ST	1.54	13		3.18	20		7.19	21	
0.44	6		1.55	14		3.19	21	ST	40 min break		
0.47	7		1.56	15	ST	3.22	12		8.00	1	
0.48	8	ST	1.59	14		3.22**	20		8.05	2	
0.49	4		2.00	1		3.23	21		8.10	3	
0.51	7		2.00**	15		3.24	19		8.19	4	
0.52	8		2.01	13		3.25	16		8.24	5	
0.53	9	ST	2.04	15		3.27	21		8.29	6	
0.54	5		2.05	2		3.29	20		8.43	7	
0.56	8		2.06	14		3.30	17		8.48	8	
0.57	9		2.10	3		3.34	21		8.53	9	
0.58	7		2.11	15		3.35	18		9.12	10	
0.59	6		2.12	10		3.39	19		9.17	11	
1.00	1		2.16	13		3.44	20		9.22	12	
1.01	9		2.17	11		3.46	13		9.46	13	
1.03	8		2.19	4		3.49	21		9.51	14	
1.05	2		2.21	14		3.51	14		9.56	15	
1.08	9		2.22	12		3.56	15		½ hr break		
1.10	3		2.24	5		4.00	1		10.25	16	
1.12	10	ST	2.25	16	ST	4.05	2		10.30	17	
			2.26	15		4.09	19		10.35	18	
			2.29	6		4.10	3		11.09	19	
						4.14	20		11.14	20	
									11.19	21	

*ST = start time (i.e., precise moment stirring ceases).

**Coincident withdrawal times: take the 2hr (or 4hr) readings as soon as convenient (an error of about 30 secs over several hours is not significant).

NO.	ET	R'h	D	$\dfrac{R'h}{C_c}$	P	NO.	ET	R'h	D	$\dfrac{R'h}{C_c}$	P	NO.	ET	R'h	D	$\dfrac{R'h}{C_c}$	P
	.01						.01						.01				
	.02						.02						.02				
	.01						.01						.01				
	.02						.02						.02				
	.01						.01						.01				
	.02						.02						.02				

Figure 64. Hydrometer worksheet. ET = elapsed time.

3. Start the timer at 1 min ahead of zero, and immediately begin stirring the column. At time zero, withdraw the stirrer. Follow the program (Table 11 for hydrometer reading times. This program gives reading times for multiple columns at 4, 8, 15, and 30 min, and 1, 2, 4, and 8 hr, corresponding to approximately 5.5, 6.0, 6.5, 7.0, 7.5, 8.0, 8.5, and 9.0$\emptyset$ size fractions, respectively. For finer sizes, take one or two readings the next day, noting the time. Cover the cylinders with watch glasses when there are any long periods between readings.

Calculations

Transfer the data from the program (Table 11) to a data sheet such as Fig. 64. Calculate $R'h - Cc$, diameter (D), and percent (P) according to the following formulas.

Corrected hydrometer reading (Rh) = $R'h - Cc$

where $R'h$ is the hydrometer reading at the top of the meniscus and Cc is the composite correction factor from the readings in the reference cylinder.

Diameter in mm (D) = $K\sqrt{H_r/t}$

where K = a constant depending on the temperature of the suspension and the solid density of the particles (see Table 12); H_r = effective depth (read

from the graph that plots $R'h$ versus H_r); and t = elapsed time after stirring (in min).

$$Percentage \; (coarser \; than) \; (P) = 100 - \frac{100 \times Rh \times a}{W}$$

where Rh = corrected hydrometer reading, a = S.G. correction (see Table 13), and W = weight of total sample.

Table 13. Specific Gravity Correction

SG	a
2.95	0.94
2.90	0.95
2.84	0.96
2.80	0.97
2.75	0.98
2.70	0.99
2.65	1.00
2.60	1.01
2.55	1.02
2.50	1.03
2.45	1.05

Table 12. Values of K for a Range of Temperatures and Solid Density of Soil Particles in Hydrometer Analysis

Temp. (°C)	Solid density of soil particles of silt and clay fraction (g/cm³)								
	2.45	2.50	2.55	2.60	2.65*	2.70	2.75	2.80	2.85
16	.00484	.00476	.00468	.00461	.00454	.00447	.00441	.00434	.00429
17	.00478	.00470	.00462	.00455	.00448	.00441	.00435	.00429	.00423
18	.00472	.00464	.00456	.00449	.00442	.00436	.00430	.00423	.00418
19	.00466	.00458	.00451	.00444	.00437	.00430	.00424	.00418	.00413
20	.00460	.00453	.00445	.00438	.00432	.00425	.00419	.00413	.00408
21	.00455	.00447	.00440	.00433	.00426	.00420	.00414	.00408	.00403
22	.00449	.00442	.00434	.00428	.00421	.00415	.00409	.00404	.00398
23	.00444	.00437	.00429	.00423	.00416	.00410	.00404	.00399	.00393
24	.00439	.00432	.00424	.00418	.00411	.00405	.00400	.00394	.00389
25	.00434	.00427	.00420	.00413	.00407	.00401	.00395	.00390	.00384
26	.00429	.00422	.00415	.00408	.00402	.00396	.00391	.00385	.00380
27	.00424	.00417	.00410	.00404	.00398	.00392	.00386	.00381	.00376
28	.00420	.00412	.00406	.00400	.00393	.00387	.00382	.00377	.00372
29	.00415	.00408	.00401	.00395	.00389	.00383	.00378	.00373	.00367
30	.00410	.00404	.00397	.00391	.00385	.00379	.00374	.00368	.00363

*Column most frequently used.

RAPID ANALYSIS FOR SAND/SILT/CLAY RATIO

A simplified but precise sieve plus pipette analysis can be made rapidly when ratios of sand to silt to clay are all that is necessary.

1. Prepare a representative subsample with 15-20g of mud.

2. Disaggregate and disperse the subsample in a solution of distilled water plus (known quantity of) dispersant.

3. Wet sieve the subsample and measure weight of dried sand fraction as in steps 3 and 4 of "Pipette Analysis for Mud." Return the pan fraction below the 4∅ dry sieve to the mud fraction.

4. Perform pipette analysis as described *but only for two subsamples:* 4∅ (for total mud) and 8∅ (for total clay).

5. Subtract weight percent of 8∅ fraction from that of 4∅ fraction to obtain total silt weight percent. Calculate weight percent of sand fraction.

LITERATURE

Sediment Textures

Baker, H. W., Jr., 1976, Environmental sensitivity of submicroscopic surface textures on quartz sand grains—statistical evaluation, *Jour. Sed. Petrology* **46**:871-880.

Barrett, P. J., 1980, The shape of rock particles, a critical review, *Sedimentology* **27**:291-303.

Bradley, W. C., 1970, Effects of weathering on abrasion of granitic gravel, Colorado River (Texas), *Geol. Soc. America Bull.* **81**:61-80.

Brewer, R., 1964, *Fabric and Mineral Analysis of Soils,* Wiley, New York. (Reprinted in 1976 by Robert E. Kreiger Publishing, Huntington, N.Y., 482p.).

Brewer, R. C., and A. D. Haldane, 1957, Preliminary experiments on the development of clay orientations in soils, *Soil Science* **84**:301-309.

Crook, K. A. W., 1968, Weathering and roundness of quartz sand grains, *Sedimentology* **11**:171-182.

Faas, R. W., and C. A. Nittrouer, 1976, Post depositional facies development in the fine grained sediments of the Wilkinson Basin, Gulf of Maine, *Jour. Sed. Petrology* **46**:337-344.

Folk, R. L., 1951, Stages of textural maturity, *Jour. Sed. Petrology* **21**:127-130.

Folk, R. L., 1954, The distinction between grain size and mineral composition in sedimentary rocks, *Jour. Geology* **62**:344-359.

Folk, R. L., 1974 (and other years), *Petrology of Sedimentary Rocks,* Hemphill, Austin, Tex., 182p.

Folk, R. L., P. B. Andrews, and D. W. Lewis, 1970, Detrital sedimentary rock classification and nomenclature for use in New Zealand, *New Zealand Jour. Geology and Geophysics* **13**:937-968.

Garrels, R. M., and F. T. Mackenzie, 1971, *Evolution of Sedimentary Rocks,* W. W. Norton & Co., New York, 397p.

Graton, L. C., and H. J. Fraser, 1935, Systematic packing of spheres—with particular relation to porosity and permeability, *Jour. Geology* **43**:785-909.

Halley, R. B., 1978, Estimating pore and cement volume in thin section, *Jour. Sed. Petrology* **48**:642-650.

Haven, D. S., and R. Morales-Alamo, 1968, Occurrence and transport of faecal pellets in suspension in a tidal estuary, *Sed. Geology* **2**:141-151.

Heezen, B., and C. Hollister, 1964, Deep-sea current evidence from abyssal sediments, *Marine Geol.* **1**:141-174.

Krinsley, D. H., and J. C. Doornkamp, 1973, *Atlas of Quartz Sand Surface Textures,* Cambridge University Press, Cambridge, 91p.

Krumbein, W. C., 1934, Size frequency distributions of sediments, *Jour. Sed. Petrology* **4**:65-77.

Krumbein, W. C., 1941, Measurement and geological significance of shape and roundness of sedimentary particles, *Jour. Sed. Petrology* **11**:64-72.

Middleton, G. V., and J. B. Southard, 1977, *Mechanics of Sediment Movement,* SEPM Short Course No. 3, Society of Economic Paleontologists and Mineralogists, Tulsa, Okla. (esp. chap. 6).

Moss, A. J., 1972, Bed load sediments, *Sedimentology* **18**:159-219.

Pettijohn, F. J., 1975, *Sedimentary Rocks,* Harper & Row, New York, 628p.

Prokopovich, N. P., 1969, Deposition of clastic sediment by clams, *Jour. Sed. Petrology* **39**:891-901.

Pryor, W. H., and W. A. Vanwie, 1971, The Sawdust Sand—an Eoecene sediment of floccule origin, *Jour. Sed. Petrology* **41**:763-769.

Rittenhouse, G., 1943, A visual method of estimating two-dimensional sphericity, *Jour. Sed. Petrology* **13**:79-81.

Rittenhouse, G., 1946, Grain roundness—a valuable geologic tool, *Am. Assoc. Petroleum Geologists Bull.* **30**:1192-1197.

Shea, J. H., 1974, Deficiencies of clastic particles of certain sizes, *Jour. Sed. Petrology* **44**:985-1003.

Shepard, F. P., and R. Young, 1961, Distinguishing between beach and dune sands, *Jour. Sed. Petrology* **31**:196-214.

Sneed, E. P., and R. L. Folk, 1958, Pebbles in the Lower Colorado River, Texas, a study in particle morphogenesis, *Jour. Geology* **66**:114-150.

Taira, A., and B. R. Lienert, 1979, The comparative reliability of magnetic, photometric, and microscopic methods of determining the orientations of sedimentary grains, *Jour. Sed. Petrology* **49**:759-772.

Tanner, W. F., 1969, The particle size scale, *Jour. Sed. Petrology* **39**:809-812.

Waddell, H., 1932, Volume, shape, and roundnes of rock particles, *Jour. Geology* **40**:443-451.

Waddell, H., 1935, Volume, shape, and roundness of rock particles, *Jour. Geology* **43**:250-280.

Wentworth, C. K., 1922, A scale of grade and class terms for clastic sediments, *Jour. Geology* **30**:377-392.

Whetton, J. T., and J. W. Hawkins, Jr., 1970, Diagenetic origin of greywacke matrix minerals, *Sedimentology* **15**:347-361. (See also discussion by J. P. B. Lovell and reply in *Sedimentology* **19**:141-146.)

Quantitative Analysis

Adams, J., 1977, Sieve size statistics from grain measurements: *Jour. Geology* **85**:209-227.

Allen, G. P., 1971, Relationship between grain size parameter distribution and current patterns in the Gironde Estuary (France), *Jour. Sed. Petrology* **41**:74-88.

Andrews, P. B., and G. J. van der Lingen, 1969, Environmentally significant characteristics of beach sands, *New Zealand Jour. Geology and Geophysics* **12**:119-137.

Baba, J., and P. D. Komar, 1981, Measurements and analysis of settling velocities of natural quartz sand grains, *Jour. Sed. Petrology* **51**:631-640.

Carroll, D. C., 1941, Grain counts with the petrographic microscope, *Jour. Sed. Petrology* **11**:44-45.

Carver, R. E. (ed.), 1971, *Procedures in Sedimentary Petrology*, Wiley-Interscience, New York, 653p.

Chayes, F., 1956, *Petrographic Modal Analysis*, Wiley, New York, 113p.

Coates, G., and C. Hulse, 1981, "Granpop" program for calculating cumulative percentages, percentiles, and grain-size parameters from sieve and pipette data using TI59 calculator and printer, *New Zealand Geol. Survey Report G53*, 23p.

Cochran, W. G., 1977, *Sampling Techniques*, Wiley, New York, 428p.

Cronan, D. S., 1972, Skewness and kurtosis in polymodal sediments from the Irish Sea, *Jour. Sed. Petrology* **42**:102-106.

Davis, M. W., and R. Erlich, 1970, Relationship between measures of sediment-size-frequency distributions and the nature of sediments, *Geol. Soc. America Bull.* **81**:3537-3548.

Emery, K. V., 1978, Grain size in laminae of beachsand, *Jour. Sed. Petrology* **48**:1203-1212.

Erlich, R., P. J. Brown, J. M. Yarus, and R. S. Przygocki, 1980, The origin of shape frequency distributions and the relationship between size and shape, *Jour. Sed. Petrology* **50**:475-484.

Folk, R. L., 1966, A review of grain-size parameters, *Sedimentology* **6**:73-93.

Folk, R. L., and W. C. Ward, 1957. Brazos River bar: a study in the significance of grain-size parameters, *Jour. Sed. Petrology* **27**:3-26.

Friedman, G. M., 1958, Determination of sieve-size distribution from thin section data for sedimentary petrological studies, *Jour. Geology* **66**:394-416.

Friedman, G. M., 1962, On sorting, sorting coefficients, and the lognormality of the grain-size distribution of sandstones, *Jour. Geology* **70**:737-753.

Friedman, G. M., 1967, Dynamic processes and statistical parameters compared for size frequency distribution of beach and river sands, *Jour. Sed. Petrology* **37**:327-354.

Friedman, G. M., 1979, Address of retiring President of the International Association of Sedimentologists: differences in size distributions of populations of particles among sand grains of various origins, *Sedimentology* **26**:3-32.

Grace, J. T., B. T. Grothaus, and R. Erlich, 1978, Size frequency distributions taken from within sand laminae, *Jour. Sed. Petrology* **48**:1193-1202.

Griffiths, J. C., 1967, *Scientific Method in Analysis of Sediments*, McGraw-Hill, New York, 508p.

Harrell, J. A., and R. A. Eriksson, 1979, Empirical conversion equations for thin section and sieve derived size distribution parameters, *Jour. Sed. Petrology* **49**:273-280.

Jones, T. A., 1970, Comparison of descriptors of sediment grain size distributions, *Jour. Sed. Petrology* **40**:1214-1215. (Also see discussion by B. K. Sahu and reply in *Jour. Sed. Petrology* **41**:1150-1153.)

Kellerhals, R., J. Shaw, and V. K. Arora, 1975, On grain size from thin section, *Jour. Geology* **83**:79-96.

Kelley, J. T., 1981, Size distribution of disaggregated inorganic suspended sediment: southern New Jersey inner continental shelf, *Jour. Sed. Petrology* **51**:1097-1101.

Klovan, J. E., 1966, The use of factor analysis in determining depositional environments from grain-size distributions, *Jour. Sed. Petrology* **36**:115-125.

Krumbein, W. C., 1935, Thin section mechanical analysis of indurated sediments, *Jour. Geology* **43**:482-496.

Krumbein, W. C., 1968, Statistical models in sedimentology, *Sedimentology* **10**:7-24.

Krumbein, W. C., and F. J. Pettijohn, 1938, *Manual of Sedimentary Petrography*, Appleton-Century-Crofts, New York, 549p.

Leroy, S. D., 1981, Grain-size and moment measures: a new look at Karl Pearson's ideas on distribution, *Jour. Sed. Petrology* **51**:625-630.

McCammon, R. B., 1976, A practical guide to the construction of bivariate scatter diagrams, *Jour. Sed. Petrology* **46**:301-304.

McLaren, P., 1981, An interpretation of trends in grain-size measures, *Jour. Sed. Petrology* **51**:616-624.

Macpherson, J. M., and D. W. Lewis, 1978, What are you sampling? *Jour. Sed. Petrology* **48**:1341-1343.

Martins, L. R., 1965, Significance of skewness and kurtosis in environmental interpretation, *Jour. Sed. Petrology* **35**:768-769.

Mason, C. C., and R. L. Folk, 1958, Differentiation of beach, dune, and eolian flat environments by size analysis, Mustang Island, Texas, *Jour. Sed. Petrology* **28**:211-226.

Middleton, G. V., 1976, Hydraulic interpretation of sand size distributions, *Jour. Geology* **84**:405-426.

Moiola, R. J., and D. Weiser, 1968, Textural parameters: an evaluation, *Jour. Sed. Petrology* **38**:45-53.

Passega, R., 1964, Grain size representation by CM patterns as a geologic tool, *Jour. Sed. Petrology* **34**:830-847.

Passega, R., 1972, Sediment sorting related to basin mobility and environment, *Am. Assoc. Petroleum Geologists Bull.* **56**:2440-2450.

Passega, R., 1977, Significance of CM diagrams of sediments deposited by suspension, *Sedimentology* **24**:723-733.

Pettijohn, F. J., 1975, *Sedimentary Rocks*, Harper & Row, New York, 628p.

Reed, W. E., R. le Fever, and G. J. Moir, 1975, Depositional environment interpretation from settling velocity (psi) distributions, *Geol. Soc. America Bull.* **86**:1321-1328.

Rosenfeld, M. A., L. Jacobsen, and J. C. Ferm, 1953, A comparison of sieve and thin section techniques for size analysis, *Jour. Geology* **61**:114-132.

Slatt, R. M., and D. E. Press, 1976, Computer program for presentation of grain size data by the graphic method, *Sedimentology* **23**:121-131.

Taira, A., and P. A. Scholle, 1979, Discrimination of depositional environments using settling tube data, *Jour. Sed. Petrology* **49**:787-800.

Tucker, R. W., and H. L. Vasher, 1980, Effectiveness of discriminating beach, dune and river sands by moments and the cumulative weight percentage, *Jour. Sed. Petrology* **50**:165-172.

Visher, G. S., 1969, Grain size distributions and depositional processes, *Jour. Sed. Petrology* **39**:1074-1106.

Warren, G., 1974, Simplified form of the Folk-Ward skewness parameter, *Jour. Sed. Petrology* **44**:259.

Pretreatment of Samples

Anderson, J. U., 1963, Effects of pretreatments on soil dispersion, *New Mexico State College Agricultural Experimental Station Research Report 78*, 14p.

Bodine, M. W., Jr., and T. H. Fernald, 1973, EDTA dissolution of gypsum, anhydrite, and Ca-Mg carbonates, *Jour. Sed. Petrology* **43**:1152-1156.

Brewer, R., 1964, *Fabric and Mineral Analysis of Soils*, Wiley, New York. (Reprinted in 1976 by Robert E. Kreiger Publishing, Huntington, N. Y., 482p.)

Drosdoff, M., and E. Truog, 1935, A method for removing iron oxide coating from minerals, *Am. Mineralogist* **20**:669-673.

Glover, E. D., 1961, Method of solution of calcareous materials using the complexing agent EDTA, *Jour. Sed. Petrology* **31**:622-626.

Jackson, M. L., 1958, *Soil Chemical Analysis*, Prentice-Hall, Englewood Cliffs, N. J., 498p.

Kaddah, M. T., 1974, The hydrometer method for detailed particle-size analysis. I. Graphical interpretation of hydrometer readings and test of method, *Soil Sci.* **118**:102-108.

Kravitz, J. H., 1966, Using an ultrasonic disrupter as an aid to wet sieving, *Jour. Sed. Petrology* **36**:811-812.

Krumbein, W. C., and F. J. Pettijohn, 1938, *Manual of Sedimentary Petrography*, Appleton-Century-Crofts, New York, 549p.

Kunze, G. W., 1965, Pretreatment for mineralogical analysis, in C. A. Black (ed.), *Methods of Soil Analysis*, vol. 1, American Society of Agronomy, Inc., Madison, Wisc., pp. 568-577.

Leith, C. J., 1950, Removal of iron oxide coatings on mineral grains, *Jour. Sed. Petrology* **20**:174-176.

Milner, H. B., 1962, *Sedimentary Petrography*, vol. 1, *Methods in Sedimentary Petrography*, Allen & Unwin, London, 643p.

Mitchell, B. D., and R. C. McKenzie, 1954, Removal of free iron oxide from clays, *Soil Sci.* **77**:173.

Moston, R. P., and A. I. Johnson, 1964, Ultrasonic dispersion of samples of sedimentary deposits, *U.S. Geol. Survey Prof. Paper 501-C*:159-160.

Mueller, G., 1967, *Methods in Sedimentary Petrology*, H.-U. Schmincke, trans., Hafner, New York, 38-39.

Prokopovich, N. P., and C. K. Nishi, 1967, Methodology of mechanical analysis of subaqueous sediments, *Jour. Sed. Petrology* **37**:96-101.

Savage, E. L., 1969, Ultrasonic disaggregation of sandstones and siltstones, *Jour. Sed. Petrology* **39**:396-398.

Truog, E., J. R. Taylor, R. W. Pearson, M. E. Weeks, and R. W. Simonson, 1937, Procedure for special type of mechanical and mineralogical soil analysis, *Soil Sci. Soc. America Proc.* **1**:101-112.

Tchillingarian, G., 1952, Study of the dispersing agents, *Jour. Sed. Petrology* **22**:229-233.

Zingula, R. P., 1968, A new breakthrough in sample washing, *Jour. Paleontology* **42**:1092.

Composition of Detrital Sediments

CLASSIFICATION SYSTEM FOR SANDSTONES

According to Lyell (1837, pp. 456–457), a sandstone is "any stone which is composed of an agglutination of grains of sand, whether calcareous, siliceous, or of any other mineral nature." In that definition, classification of a rock as a sandstone is independent of the mineralogical composition of its grains. Many limestones and some other sediments with components formed in the depositional environment have a texture of sand-size particles.

It is convenient to have a nomenclature that distinguishes between the texture of a rock and the composition of its grains. Thus, while the term *sandstone* indicates texture, the term *arenite* (from Latin *arena*, sand) is used to refer to the composition of a detrital sandstone, that is, a sandstone with grains comprising detritus from other rocks. *Arenite* is used in this way regardless of whether the sample is soft (sand) or indurated (sandstone); no confusion will result.

The use of *arenite* should be restricted theoretically to sand-size grains (2.0–0.06 mm in diameter). However, for purposes of compositional classification, extension of the limits through coarse silt at one end and granules at the other is both convenient and practical. Special techniques are generally necessary to determine the composition of grains smaller than about 0.03 mm (the lower limit of coarse silt). Thus detrital materials smaller than this size are relegated to the status of *matrix*. Matrix is taken into account in the textural classification and is ignored in the compositional classification (although its composition must be mentioned in the sediment description). The composition of particles of pebble or larger sizes cannot be evaluated representatively in thin-section analysis, the most common technique used in studying arenites. Thus grains larger than 4 mm (the upper limit for granules) are best excluded from an arenite classification. Purists may wish to exclude both coarse silt and granules.

Arenite refers only to the *detrital* fraction of a sandstone. Sand grains of biogenic, biochemical, and chemical origin are not included. Rocks that contain more than 50% *non*detrital grains by volume are included in separate classifications, but it may be appropriate to give both a detrital and nondetrital name to a sediment with substantial components of each. The presence of smaller quantities of nondetrital grains may be indicated by adding the name of the grain type as an adjectival modifier (for example, *fossiliferous*) to the arenite name. Postdepositional precipitates that act as *cements* also must be excluded when determining the basic arenite name. They should be noted as another modifier of the arenite name, for example, carbonate-cemented, calcite-cemented, iron-oxide-cemented, pyrite-cemented.

Klein (1963) and McBride (1963) have reviewed some of the existing classifications for describing detrital sandstone composition and have pointed out deficiencies in many of them. Convergent trends are apparent in the organization of the modern classifications, both in these reviews and in more recent papers dealing with detrital sandstones. No single classification is in general use, however. Many redefine older terms, which therefore vary widely in connotation. The classification for detrital sandstone composition presented here is in accordance with recent trends, employs terminology that is self-explanatory, and avoids terms that have been extensively redefined in previous schemes, such as arkose (for example, see Pettijohn, 1943; Oriel, 1949; Folk, 1974) and graywacke (see Boswell, 1960;

Dott, 1964). Nevertheless, in any formal report it is important that you clearly reference the system of classification that you are using. This scheme has been formally published by Folk, Andrews, and Lewis (1970).

Most detrital sand grains are either quartz (Q), feldspar (F), or rock fragments (R). Thus, a ternary diagram is constructed in which the QFR components of any arenite may be quantitatively plotted (Fig. 65).

Standardization is necessary in assigning mineral grains to the poles of the QFR triangle. At the Q pole are grouped all monocrystalline and polycrystalline quartz grains excluding clastic chert (a rock fragment). While all polycrystalline grains might be considered rock fragments, practical difficulties at present preclude consistent distinction between plutonic polycrystalline quartz, sedimentary quartzite, and metaquartzite fragments (see, for instance, Blatt and Christie, 1963; Blatt, 1967a, and 1967b, and Fig. 72). In addition, it is generally impossible to distinguish monocrystalline quartz grains from these different sources. Thus it is unreasonable to assign monomineralic quartz grains to different poles of the triangle. (Quartzite grains with mica are polymineralic and may be assigned to the R pole.)

At the F pole are grouped all monocrystalline feldspars. At the R pole are grouped all recognizable rock fragments (igneous, metamorphic, and sedimentary). Granite and gneiss grains are assigned to the R pole in this classification, an innovation on most previous systems, wherein these components were lumped with feldspars (because most previous workers assumed that most feldspars were derived from granitic rocks, an assumption that is not valid for many). Polycrystalline, monomineralic feldspars are also placed in the R category—the size of the component crystals usually permits identification of these grains as of either volcanic (small lathlike crystals) or plutonic (coarse anhedral crystals) derivation.

Subdivision of the QFR triangle is largely arbitrary. The class limits are selected along boundaries that are easy to remember and they delineate fields that can be simply and meaningfully named. Classes

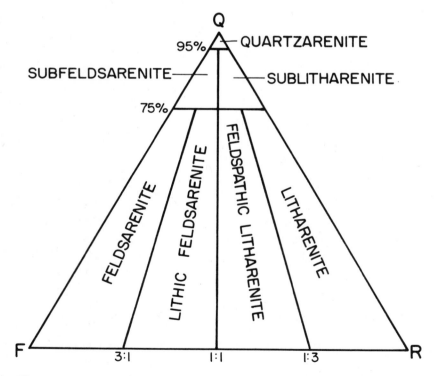

Figure 65. Parent triangle for arenite classification. Q = monocrystalline and polycrystalline quartz (excluding chert); F = monocrystalline feldspar; R = rock fragments (igneous, metamorphic, and sedimentary, including chert). *(After Folk, Andrews, and Lewis, 1970)*

also reflect an attempt to group arenites with similar source rock lithologies (or genetic history) and to separate arenites with dissimilar source rock lithologies (or genetic history). In addition, grouping is attempted, at least partly, according to the most common geological occurrence of arenite varieties. For example, many arenites, especially in continental regions, are predominantly composed of quartz, either because of prolonged weathering that results in elimination of unstable minerals prior to deposition, or because most of the grains are recycled from other quartz-rich sedimentary (or metasedimentary) rocks (see Blatt, 1967a).

The virtually pure *quartzarenites* are arbitrarily distinguished from other quartz-rich arenites by a boundary at 95% *Q*. It is useful to report whether rock fragments are more or less common than feldspars; thus the remainder of the triangle is divided into two equal parts at a ratio of 1:1 *F* to *R* grains. It is also useful to distinguish arenites with intermediate abundances of the relatively unstable *F* and *R* grains from those with abundant unstable components (the *feldsarenites* and the *litharenites*). Previous petrological studies of arenites have shown that a boundary of 25% *F* grains is a reasonable limit for these rocks of intermediate composition. Although there is no apparent "natural" grouping of intermediate-*R* arenites, it is convenient to take the same limit for these arenites as for the intermediate-*F* arenites. Thus two fields are established subordinate to quartzarenite with a lower boundary at 75%—*subfeldsarenite* and *sublitharenite*.

At this level of subdivision, the *QFR* triangle has five fields; the two fields for arenites rich in *F* and *R* grains span very wide compositional ranges. There is too little data on which to base a subdivision of these two fields into "natural" groups. Yet subdivision would be useful in attempts to describe compositions more precisely. Purely as a choice of convenience, they are split into equal halves—that is, at ratios of 3:1 and 1:3 *F* to *R*. The two subdivisions intermediate in their compositions of *F* and *R* components are named *feldspathic litharenite* and *lithic feldsarenite*.

The geologically less common arenites that have a predominance of other types of detrital mineral grains are best treated as "(mineral)-arenites," for example, magnetite-arenite, mica-arenite. *The QFR triangle is bypassed in naming these arenites.* If these minerals (that is, minerals that are not quartz, feldspar, or rock fragments) constitute less than the *QFR* components by volume, they are ignored

in compiling a *QFR* name but appear as a preceding term, for example, mica quartzarenite. (Prominent nondetrital grains of sand size, such as glauconite and fossil fragments, are treated in the same way, but adjectival modifiers should be used. Thus one may find glauconitic quartzarenites or fossiliferous mica quartzarenites.) No limits are defined for the percentage at which these nonessential minerals should be included in the name; workers may decide for themselves.

For plotting sample compositions onto the *QFR* triangle, the percentage of the *Q*, *F*, and *R* components alone are recalculated to 100%. (Percentages of components are precisely determined by point-counting techniques but you can estimate them with reasonable precision after some experience with the aid of a comparison chart, such as Fig. 66.) For example, let us consider an arenite with a measured composition of 45% quartz, 18% feldspar, 5% rock fragments, 3% mica, 15% glauconite, and 14% carbonate cement. First calculate the *Q* component, which is $100 \times$ [quartz/(quartz + feldspar + rock fragments)] $= 100 \times$ [45/(45 + 18 + 5)] = 66.2%. The point representing the composition of this arenite would thus lie on the line of 66.2% *Q* (parallel to the base of the triangle and two-thirds of the way toward the *Q* apex). (The *F*:*R* ratio is 3.6:1; thus the point lies in the feldsarenite field, and you have the compositional *name* without proceeding any further.) Now recalculate the *F* and/or *R* components. The *F* component is ($100 \times 18/68$), or 26.5%, and the *R* component is ($100 \times 5/68$), or 7.3%. The sample point lies at the intersection of the 7.3% *R* line (parallel to the *QF* side of the triangle) or of the 26.5% *F* line with the 66.2% *Q* line. (If the three percentages do not intersect at the same point, you have made an error!) An asterisk marks the location of the example in Fig. 65.

With regard to the outlined arenite scheme, a problem of particular relevance to tectonically superactive areas, or any region supplied with detritus from rocks with a paucity of quartz, lies in the abundance of arenites with extremely low quartz content. Although quantitative plotting of data on the primary triangle adequately locates quartz-poor arenites, a worker may find it useful to give them a distinctive name. For this essentially qualitative purpose, the terms *quartz-poor* or *hyper-* could be prefixed to *feldsarenite* or *litharenite*. Until sufficient data accumulate to define the most useful boundary for formal categories, you should note at what *Q*-percentage you begin to apply these names.

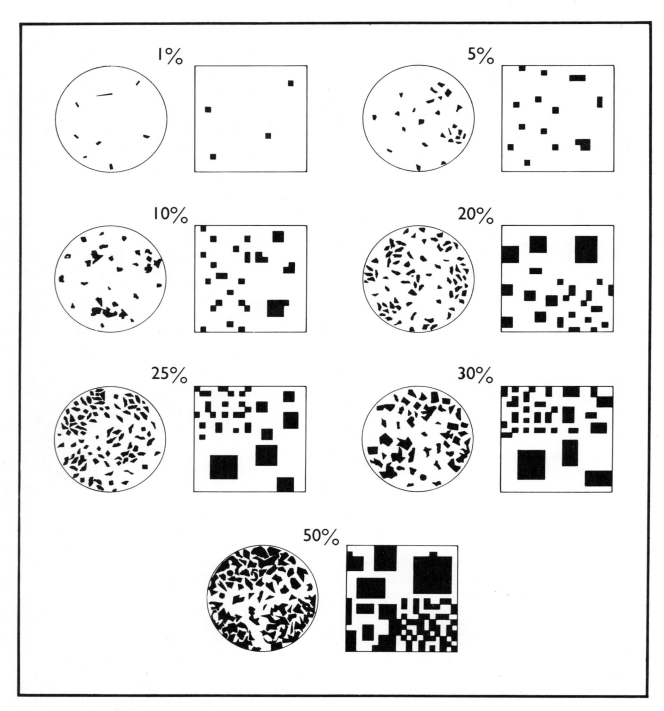

Figure 66. Percentage estimation comparison charts. *(Reproduced from Folk, Andrews, and Lewis, 1970, New Zealand Journal of Geology and Geophysics)*

Refinement of Nomenclature

For detailed analysis of arenites, refinement of the litharenite classes is generally necessary—the proportion of igneous/metamorphic/sedimentary rock fragments, or the proportion of specific varieties of these rocks fragments, are important for interpreting provenance and sediment history. In some cases, it is also useful to refine the quartz-arenite and feldsarenite classes. The degree of refinement will depend on the particular suite of arenites that you are studying.

Refinement in nomenclature for the litharenites is easily achieved by appending the name of the most common rock fragment type to -*litharenite* or -*sublitharenite* (or by substituting a contraction for *lith-*). In the case of lithic feldsarenite, replacement of the word *lithic* by the appropriate name seems reasonable. If sedimentary rock fragments predominate, the name used would be *sedimentary-litharenite* (or *sedarenite*). An *igneous-litharenite* could be named either *volcanic-litharenite* (*volcarenite*) or *plutonic-litharenite* (*plutarenite*), depending on which class of components predominate. Because of the difficulty in distinguishing

between gneiss and granite fragments in thin section (Boggs, 1968), it must be realized that plutonic litharenites would include rock fragments derived from gneisses as well as from deep-seated igneous rocks. If the phyllitic or micaceous metamorphic rock fragments predominate, *phyllarenite* is probably the most unequivocal name.

For convenience in plotting quantitative data or in comparing rock suites, second- and even third-order triangles may be devised as in Fig. 67. Any worker may select any end members that are useful for the particular suite of sediments. For example, at the sedimentary-litharenite pole of a secondary triangle, the third-order poles might be *sandstone* and/or *shale, chert,* and *carbonate* rock fragments. Note that in a situation where carbonate rock fragments predominate, the term *carbonate-litharenite* or *calclitharenite* must be used instead of *calcarenite*. Calcarenite has the pre-emptive connotation of a rock composed mainly of carbonate sand particles that have been derived from within the basin of deposition and not from the erosion of pre-existing carbonate rocks.

At the *F* pole of the primary triangle, a second-order triangle could be devised with a 50% divi-

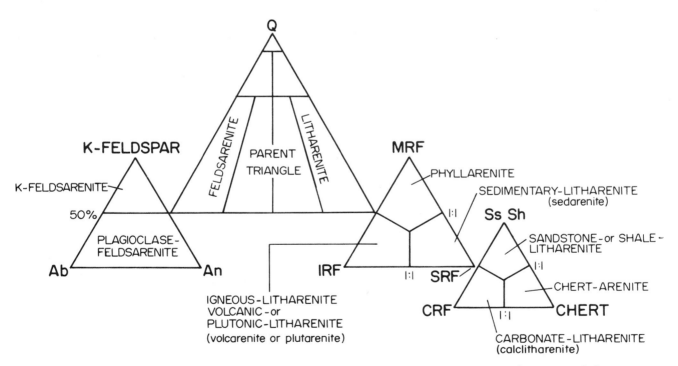

Figure 67. Examples of daughter triangles for arenite classification. *MRF* = metamorphic rock fragments; *IRF* = igneous rock fragments; *SRF* = sedimentary rock fragments; *CRF* = carbonate rock fragments. *(After Folk, Andrews, and Lewis, 1970)*

113

sion between *K-feldsarenite* and *plagioclase-felds-arenite (plagarenite)*. At the Q pole, it may be useful to devise a second-order triangle with poles for monocrystalline quartz of undulatory extinction, monocrystalline quartz of straight extinction, and polycrystalline quartz. It appears that polycrystalline quartz, then quartz of undulatory extinction, are preferentially eliminated relative to quartz of straight extinction by prolonged action of sedimentary processes.

In all cases of plotting on a second- or third-order triangle, the constituents represented by the poles of the triangle are recalculated to 100% before plotting. The main problem that arises is that points appear on two (the primary and secondary) or more triangles for any one arenite sample. Such multiple plots become cumbersome with a multitude of samples, and the worker may choose to devise some other form of comparing suites or representing composition in the detailed subdivisions.

Complete Classification

The preceding discussion has concerned composition; earlier sections have discussed texture, sedimentary structures, and, briefly, both color and induration. The complete name of an arenite (or any other sediment) should list all these properties: (color) (induration) (internal sedimentary structures) (sorting term) (size term): (cement) (prominent nondetrital grain type) (prominent detrital nonessential grain type) (compositional term).

An example of a name that may result (for an arenite) is pale greenish-gray friable, cross-laminated, poorly sorted bimodal medium and fine sandstone: calcite-cemented glauconitic mica subfeldsarenite. Such a classification may seem so lengthy as to be cumbersome, but it does provide the most distinctive and important characteristics of a deposit for both description and interpretation. Such lengthy names will also be uncommon.

The sequence of terms does not comprehensively describe a rock, and further description (such as details of the cross-lamination, cementation, types of feldspar, roundness of grain types) will always be necessary. In some cases composition may be difficult to ascertain in the field (as with many "graywackes"—a name that undoubtedly will continue to be used in the field for arenites with obscure texture and composition); completion of the name should then await laboratory study.

CLASSIFICATION OF CONGLOMERATES AND BRECCIAS

The nomenclature system for the textural group of conglomerates and gravels should parallel that for sandstones and sands, thus the compositional term for detrital varieties is *rudite* (from Latin *rudus*, rubble). Insufficient detailed compositional studies of rudites have been made to permit construction of a classification as detailed as for arenites.

For a qualitative description, it would seem reasonable to append the most common mineral or rock fragment name to the suffix -*rudite*. A number of names may be presented, with the convention that the one nearest the suffix is the most common, for example, *chert-granite-rudite*. Cements and nondetrital grains can be treated the same as they are in the arenite classification system.

For quantitative plots, triangular diagrams would seem again to be best because they are the easiest to construct. Rudites are predominantly composed of quartz, sedimentary, igneous, and/or metamorphic rock fragments. Because there are four common compositional "end-members," simple diagrams for quantitative plotting are difficult to devise. A tetrahedron could be imagined, from which triangular "slices" are taken at stated values of one component (Fig. 68, upper). In this case, class boundaries at ratios of 1:1 between end-members are probably best until sufficient data are obtained to delimit more practical boundaries. Second- and third-order triangles could be constructed from the poles of the triangular slices. Alternatively, the scheme of Pettijohn (1975) could be used (Fig. 68, lower), although it has the drawbacks of combining both textural and compositional parameters in the definition of his extraformational varieties, and of requiring genetic decisions rather than being a purely descriptive scheme.

The simple rudite classification of appending the name of the predominant component(s) as a prefix to -*rudite* is unsatisfactory where many different igneous, metamorphic, or sedimentary rock fragments are present. In such instances names such as *ignirudite, plutarudite, volcorudite, phyllarudite*, and *sedrudite* might be better. (The simpler alternatives *igneous-rudite, metamorphic-rudite*, and the like are not applicable because they imply a mode of origin rather than clast composition.) Contractions for other compositions may be useful. (*Note:* the term *calcirudite* should be avoided for

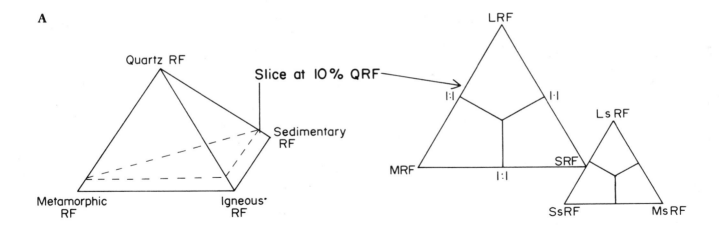

B				
EPICLASTIC	EXTRAFORMATIONAL (detrital clasts from outside depositional basin)	ORTHOCONGLOMERATES (matrix <15%)	Metastable clasts <10%	ORTHOQUARZITIC (OLIGOMICT) CONGLOMERATE (pebbles of 1 type only)
			Metastable clasts >10%	PETROMICT CONGLOMERATE (Specify dominant clast type - e.g. petromict limestone cgl.)
		PARACONGLOMERATES (matrix >15%) also termed diamictites	laminated matrix	LAMINATED CONGLOMERATIC MUDSTONE OR ARGILLITE
			non laminated matrix	TILLITE (glacial)
				TILLOID (not glacial) - pebbly mudstones, olistostromes etc
	INTRAFORMATIONAL (clasts from sediments within the depositional basin)	INTRAFROMATIONAL CONGLOMERATES AND BRECCIAS (specify dominant clast type)		
PYROCLASTIC	VOLCANIC BRECCIAS (derived from previously deposited volcanics) VOLCANIC AGGLOMERATES (formed of lava solidifed in flight)			
CATACLASTIC	LANDSLIDE AND SLUMP BRECCIAS (not involving fluid-like flow, which generates tilloids)			
	FAULT AND FOLD BRECCIAS (i.e. direct in situ results of tectonism)			
	COLLAPSE AND SOLUTION BRECCIAS (resulting from solution of underlying material - salt, limestone, -?)			
METEORIC	IMPACT OR FALLBACK BRECCIAS			

Figure 68. Alternative schemes for conglomerate: rudite classification. *A:* Example of a possible compositional classification system for a rudite suite with 10% or less quartz rock fragments. Any triangular slice from the tetrahedron could be used for quantitative plots, and daughter triangles constructed as in the example. Nomenclature (as suggested in text) would follow the textural name of the conglomerate (or breccia). *B:* Classification of conglomerates and breccias after Pettijohn (1975).

detrital carbonate rocks; as with *calcarenite* and *calcilutite*, it is widely used for rocks whose particles have been formed within the basin of deposition; *calclithrudite* is preferable.) The auxiliary term *polymictic* (see Pettijohn, 1975) may also prove useful if clasts of more than one lithologic type are present. The connotation of the term would depend on the final *-rudite* name. For example, *polymictic ignirudite* would imply that various igneous rock types predominate among the clasts, but that clasts of metamorphic and/or sedimentary rocks are also present.

Gravel-size fragments of muddy sediments are commonly generated within a basin of deposition by erosion of semiconsolidated deposits. These fragments may be termed *penecontemporaneous* or *perigenic* (implying local origin and transportation, see Lewis, 1962). The term *intraclast* is sometimes used in this way, but has the preemptive connotation of a locally-derived carbonate clast. In addition, the widely used term *intraformational* is less acceptable because to many it implies mass movement rather than fluid flow during emplacement of the clasts.

CLASSIFICATION OF MUDROCKS

To parallel the sandstone:arenite, conglomerate:rudite systems, it is appropriate to have a mudstone:lutite nomenclature (from Latin *lutum*, mud). Insufficient compositional studies of lutites have been made to permit the construction of a detailed classification system.

For qualitative description, it would seem appropriate to append the component mineral names to the suffix *-lutite*, with the convention that the name nearest the suffix is the most common—for example, *clay quartz-lutite, quartz illite-lutite*, and so on. Cements and nondetrital components can be treated as they are with arenites, although it may be difficult to distinguish authigenic from detrital clays in the rocks.

For quantitative plotting, triangular diagrams are again easiest to construct. Most lutites are predominantly composed of quartz, feldspar, and clay minerals (see Shaw and Weaver, 1965). A primary triangle with these three poles can be subdivided into three fields with boundaries at ratios of 1:1 between each component. Secondary, and even tertiary, triangles may be derived from each apex to suit the individual worker's requirements; for example, at the clay apex, a secondary triangle

could be formed with end-members of illite, montmorillonite, kaolinite. Quantitative plotting will generally require determination of composition by specialized laboratory analysis (x-ray diffraction, for example).

Note that the term *shale* is widely used for mudrocks (generally excluding siltstones; see Tourtelot, 1960); implications of fissility (a secondary feature) confuse its application and render it unsatisfactory when the primary sedimentary characteristics of the rock are being described. The term *argillite* is even worse—it refers to a mudstone or claystone that is intermediate between a shale and a slate. See Potter, Maynard, and Pryor (1980) for a general review of knowledge on mudrock sedimentology.

PETROGRAPHIC SUBDIVISION OF SOME COMMON ARENITE COMPONENTS

Quartz

Because quartz is the most common detrital mineral in arenites and because it is virtually the only one in mineralogically mature arenites, considerable attention has focused on criteria useful in interpreting subvarieties (see, for example, Blatt and Christie, 1963; Greensmith, 1963; Conolly, 1965; Moss, 1966; Blatt, 1967b; Folk, 1974; Basu et al., 1975). Some of the salient guidelines are summarized below.

DESCRIPTIVE VARIETIES

Unstrained: Grains are single crystals that extinguish as a unit when using a polarizing microscope under crossed nicols (UXN).

Strained: Grains are single crystals that never wholly extinguish UXN. An irregular band of extinction migrates across the crystal as the stage is rotated. Most grains that require rotation of the microscope stage for more than 5° to obtain extinction in different parts of the grain are probably derived from low-rank metamorphic rocks; those requiring less than 5° are probably derived from plutonic rocks (Basu et al., 1975). However, care must be taken to ensure strain did not develop after deposition with fold or fault deformation.

Polycrystalline: single grains composed of more than one crystal, each of which is strained or unstrained. Crystal boundaries may be obscure (due to recrystallization), but extinction characteristics will generally indicate the polycrystalline character. Folk (1974) classes grains with obscure crystal boundaries and similar optic orientations in the crystals as "semi-composite." Individual crystal sizes increase systematically with an increase in the source rock's metamorphic grade. More polycrystalline quartz grains in the medium sand range are initially supplied from low-rank metamorphic rocks (around 50%) than from middle- and upper-rank metamorphic (about 30%) or from plutonic rocks (about 15%) (Basu et al., 1975). The number of polycrystalline grains is also a function of grain size—the finer the size, the fewer the polycrystalline grains.

Inclusions: may be present in any quartz grain variety. These consist of vacuoles (minute bubbles, often partly filled with water; difficult to resolve even under high power) or microlites (minute crystals of other minerals, such as rutile needles). Microlite varieties may characterize quartz from specific source rocks. (See, for example, Gilligan, 1919; and Keller and Littlefield, 1950.)

GENETIC VARIETIES

Volcanic: generally water-clear, unstrained, mostly monocrystalline. Hexagonal-bipyramidal (β-quartz) shape is definitive (Fig. 69). Embayments are common due to corrosion prior to extrusion; conchoidal fractures are common; shards are rare.

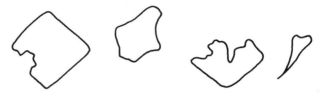

Figure 69.   Volcanic quartz.

Vein: generally abundant vacuoles (which account for the milky appearance of large grains), unstrained or slightly strained. Microlites are very rare. Vein varieties may be monocrystalline or polycrystalline (commonly with obscure crystal boundaries). They may show comb structure (appressed crystals that grew perpendicular to the vein walls) and/or contain vermicular chlorite microlites. See Fig. 70.

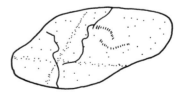

Figure 70.   Vein quartz.

Recycled sedimentary: may dominate in almost any sediment, but most grains will reflect the characteristics of the ultimate source rather than of the immediate sedimentary source. The presence of rounded or worn overgrowths is the main direct indicator. If only a few grains have overgrowths, and these do not interlock with other overgrowths, they are probably recycled (Fig. 71). (Because the overgrowths are generally in optical continuity with

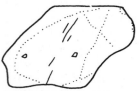
presence of overgrowths within grains is conclusive evidence

perfectly rounded grains are strong-ly suggestive

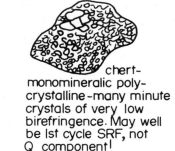
chert- monomineralic poly- crystalline-many minute crystals of very low birefringence. May well be 1st cycle SRF, not Q component!

Figure 71.   Recycled sedimentary quartz.

the parent, concentrations of minute dust along the margin of the parent crystal are commonly the only distinguishing characteristic.) Microlites will be restricted to the parent, but vacuoles may be present in both parent and overgrowth; strain may be duplicated in the overgrowth. (Rarely, sedimentary textures are apparent in polycrystalline monomineralic grains—these should be classed as SRFs. Chert clasts are an example.) Both polycrystalline and strained quartz varieties are destroyed by abrasion more rapidly than unstrained monocrystalline quartz; hence these varieties become progressively more rare as the sediment is subject to greater recycling.

Metamorphic: Often indistinguishable from primary plutonic igneous quartz (see "common" quartz). Some polycrystalline types have definitive micas (generally parallel remnants of schistosity); these are polymineralic and can be referred to as MRFs. Polycrystalline medium sand grains with 10 or more individual crystals appear to be largely of meta-

morphic origin (Fig. 72), but when similar numbers of crystals occur in finer sand sizes, be wary: they may be silicious rock fragments (see Fig. 73). The most commonly recognizable quartz grains attributable to metamorphic origin are likely to be *stretched metamorphic*, which are polycrystalline with elongate crystals showing smooth, crenulated, or granulated boundaries. In these, crystals are generally strained and have a subparallel optic orientation; microlites (especially of metamorphic minerals) may be present. They can be classed as MRFs if desirable.

Common: The vast majority of quartz grains do not have any clear characteristic indicative of their origin. Most are probably plutonic igneous/ metamorphic. They are either strained or unstrained, mono- or polycrystalline, generally with few scattered vacuoles and/or microlites. Polycrystalline varieties generally have very inequant individual crystals.

if unit crystals elongate and boundaries commonly crenulate, <u>probably</u> of metamorphic origin but safer to record as polycrystalline Q component

if with micas, MRF not Q component!

Figure 72.   Metamorphic quartz.

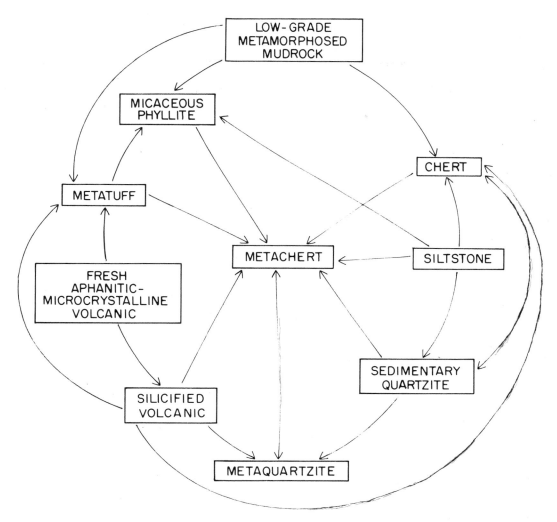

Figure 73. Textural transitions that obscure discrimination in quartzose sand and gravel grains. *(After Wolf, 1971)*

Feldspars

Feldspar varieties can prove very useful for provenance studies (see Fig. 74), and their relative degree of alteration is indicative of weathering and diagenetic history. Van der Plas (1966) provides a thorough discussion of methodology for studying feldspars in sediments; staining procedures useful for gross distinction in hand specimen or thin section are given later in this chapter.

Altered: a general category for all feldspars that have been so extensively altered that their original composition is indeterminate. Three types of feldspar alteration are commonly distinguished:

1. *vacuolization*—a multitude of minute bubbles, often partly filled with water, probably a result of hydration/hydrolysis;

2. *kaolinization*—very low birefringent microflakes)—generally too fine to distinguish; at low magnifications may resemble vacuolization;

3. *sericitization* (strictly should be "illitization" because sericite tends to be a late diagenetic or metamorphic product)—an alteration to clayey microflakes that are generally discernible by first-order white to yellow birefringence.

These alteration products may be disseminated throughout the feldspars or concentrated along cleavages; authigenic feldspar overgrowths will not show them unless alteration is late-stage diagenetic or due to late weathering of the sedimentary rock.

Microcline: shows characteristic cross-hatch twinning. (Anorthoclase, with a finer pattern, also may be distinguished.) Microcline feldspars are

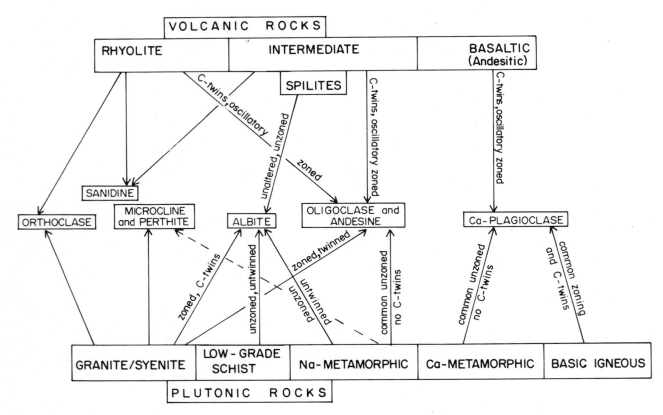

Figure 74. Feldspar provenance. *(After Pittman, 1970, and D. Shelley, University of Canterbury, pers. comm.)*

commonly fresh, and often show overgrowths (often by untwinned K-feldspar).

Orthoclase: is generally untwinned; refractive index (RI) is less than quartz and balsam; biaxial optical figure; grains are often slightly altered (unlike quartz) and often show overgrowths in sorted, originally permeable arenites. Where quartz and feldspar grains are appressed (that is, without intervening clays, carbonates, and so on), their differential relief will permit distinction under low to medium magnifications when the substage diaphragm is closed and the field of view is slightly out of focus (that is, when Becke lines are rendered obvious). Staining may be the quickest and easiest way to estimate the abundance of K-feldspars in sediments with clayey matrix, carbonate, or other cements, but note that clays may also take up the stain.

Plagioclase: is generally identified by obvious polysynthetic twinning; RI of calcium varieties is greater than that of quartz and balsam. The main problem is in distinguishing untwinned varieties of sodium-rich plagioclase (albite and oligoclase)

from quartz and orthoclase. The RI of albite is only slightly lower than that of quartz; the RI of oligoclase overlaps that of quartz, and untwinned varieties are common in metamorphic rocks. Staining should identify oligoclase, but some albite has so little calcium that it may not stain well (or potassium-bearing clayey alteration products may collect the stain as for K-feldspar).

Rock Fragments

Rock fragments (see discussion of litharenites) are the best indicators of provenance and should be carefully sought in thin section (see Table 14). Any polymineralic grain and some polycrystalline monomineralic grains (such as chert and feldspars) are rock fragments. (Grains with small microlites as inclusions of one mineral in another, which originated during magmatic crystallization or metamorphic recrystallization, do not normally count as rock fragments.) Identification of rock fragments is dependent on the size of the sand grains as well as the character of the source rock (see Fig. 75).

Table 14. Provenance Indicators

Sedimentary Provenance

mudstone/sandstone/limestone fragments
chert
quartz (particularly if abraded
    overgrowths)
altered feldspars (rounded)
abraded glauconite

zircon
rutile
tourmaline
sphene
} particularly if
rounded

other rounded hard/tough heavy
    minerals
leucoxene (?)

Low-Rank Metamorphic Provenance

slate/phyllite/quartzite fragments
muscovite
chlorite
quartz (especially metaquartzite types)
altered feldspars

leucoxene
tourmaline (particularly small
    euhedral brown xls)
other stable HMs (as for
    sedimentary source)

Higher-Rank Metamorphic Provenance

schist/gneiss fragments
muscovite/biotite
chlorite
feldspars
quartz (especially metaquartzite types)

garnet
epidote/zoisite
staurolite
kyanite/sillimanite/andalusite
magnetite/ilmenite
sphene
zircon

Acid Igneous

acid volcanics or granitic/syenitic
    fragments
quartz (common or volcanic types)
K-feldspar
Na-plagioclase
biotite/muscovite
hornblende

monazite/sphene/rutile
tourmaline
zircon
magnetite
apatite

Basic/Intermediate Igneous

basic volcanic (less commonly plutonic)
    fragments
olivine (often serpentinized)
calcic plagioclase
pyroxenes

ilmenite/magnetite
anatase/brookite
rutile
chromite

Pegmatite

quartz (common type)
orthoclase
microcline
Na-plagioclase
muscovite

cassiterite
tourmaline
beryl
topaz
monazite
fluorite
(other uncommon ones)

Note: Only rock fragments are conclusive by themselves; in general it will
be the *suite* of minerals present that will be indicative. This table has been compiled
from various sources (for example, Pettijohn, 1975).

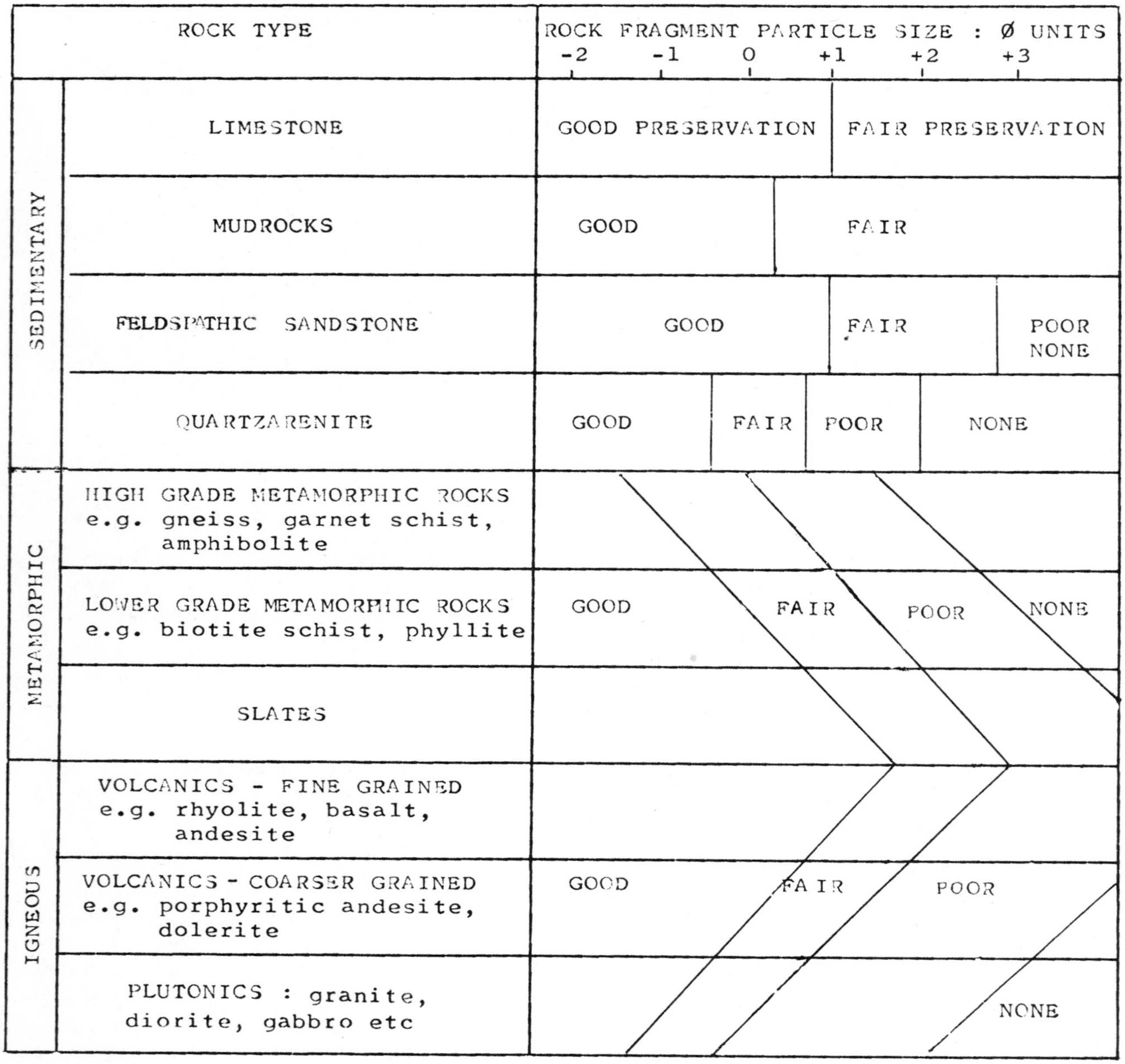

ROCK TYPE	ROCK FRAGMENT PARTICLE SIZE : Ø UNITS −2 −1 0 +1 +2 +3		
SEDIMENTARY — LIMESTONE	GOOD PRESERVATION	FAIR PRESERVATION	
MUDROCKS	GOOD	FAIR	
FELDSPATHIC SANDSTONE	GOOD	FAIR	POOR NONE
QUARTZARENITE	GOOD	FAIR POOR	NONE
METAMORPHIC — HIGH GRADE METAMORPHIC ROCKS e.g. gneiss, garnet schist, amphibolite			
LOWER GRADE METAMORPHIC ROCKS e.g. biotite schist, phyllite	GOOD	FAIR POOR	NONE
SLATES			
IGNEOUS — VOLCANICS - FINE GRAINED e.g. rhyolite, basalt, andesite			
VOLCANICS - COARSER GRAINED e.g. porphyritic andesite, dolerite	GOOD	FAIR POOR	
PLUTONICS : granite, diorite, gabbro etc			NONE

Figure 75. Relative degree of preservation of parent-rock textures shown as a function of rock fragment size. *(After Boggs, 1968)*

The main problem in identifying rock fragments lies in distinguishing between the very fine-grained/finely crystalline varieties, among which there are textural transitions (see Fig. 73). No hard and fast rule can be generally applied; indications as to the probable character of the fragments may come from other rock fragments and minerals in the sediment or a knowledge of the general geological setting and/or likely provenance.

In plane-polarized light (PPL), all may appear colorless or a murky pale brown, but PPL views are essential to distinguish grain outlines where muddy or cherty matrix is (or appears to be) present. Under Crossed Nicols, chert appears as a mosaic of generally equal-sized anhedral crystals smaller than 20 μ with very low birefringence (grays). With diagenetic crystal growth, it can develop crystals of different sizes and become virtually indistinguishable from other siliceous rock fragments. Siltstone clasts have quartz crystals up to 62.5 μ that are most commonly angular but do not form a good mosaic texture unless overgrowth interlocking or metamorphic recrystallization has occurred. Volcanic clasts commonly contain some lathlike feldspars and exhibit a porphyritic texture; shard outlines may be present in tuffs. Foliate fabric and presence of abundant phyllosilicates are diagnostic of the metamorphosed mudrocks.

Heavy Minerals

Heavy minerals (HMs) are generally sparse in thin sections from sedimentary rocks, but they may be important in determining provenance (see Table 14) and aspects of the history of the deposit (see later discussion). Hence they should be identified as accurately as is warranted by the nature of the study. Standard optical mineralogy texts or books specializing in heavy minerals should be consulted (for example, Baker, 1962; Milner, 1962; Parfenoff et al., 1970). Note that crystal habit is commonly useful, particularly with authigenic minerals and where grains are too small for good optical figures. It may prove worthwhile to recognize several varieties of specific minerals (such as zircon and tourmaline) when undertaking detailed provenance (or diagenetic) studies in a particular petrographic province — varieties based on color, habit, inclusions, or other distinctive characteristics you can select for yourself. Trends or changes of significance may become apparent when you synthesize the data with other information.

Opaque minerals should be distinguished from translucent HMs. Although the mineralogy of most opaques cannot be determined in thin section, some distinction can be made by illuminating the top of the section with a microscope lamp and studying the character of the light reflected from the mineral.

Leucoxene (a mixture of Ti-oxides and hydroxides): indicated by a bright, white reflection (do not confuse it with the adamantine luster of some translucent heavies). A common diagenetic alteration product of titanium-bearing minerals (such as ilmenite and rutile),it may recrystallize to form tiny euhedral crystals of anatase, brookite, or sphene.

Hematite and limonite (a mixture of Fe-oxides and hydroxides): indicated by red and reddish to yellowish brown reflected colors. These are common weathering and diagenetic alteration products of iron-bearing minerals. A small amount may produce marked coloration and obscure the optical properties of grains (for example, hematite in altered biotite flakes). Hematite is most commonly present as microspecks (distinguishable only under high magnification), and its distribution relative to grain overgrowths and alteration products commonly indicates whether it is a weathering or a diagenetic product. Limonite appears to be mobile in the interstitial fluids during diagenesis and may appear as discontinuous stringers or continuous bands (such as Leisegang bands) in the rock; it commonly is an obscuring stain of other minerals (such as carbonates, where it may have been exsolved during the conversion of siderite to calcite).

Pyrite and marcasite: give a brassy yellow, metallic reflection. Distinction between the sulfides in thin section must be based on crystal form. Both commonly show partial oxidation to limonite.

Magnetite, ilmenite, and other opaque minerals: cannot easily be distinguished with reflected light in thin section. Many opaque minerals give a similar metallic reflection.

Carbonaceous (and bituminous) matter: nonreflective, generally dull brown to black in reflected light (vitrinite is a shiny exception).

Matrix and Cement

Mineral matter occupying interstitial spaces between framework grains in sediments is either *matrix* — clastic material deposited mechanically at the time of sedimentation — or *cement* — material chemically precipitated from solution after deposition. Note that these definitions are *genetic*. Objective distinction is generally feasible: cement comprises a visibly crystalline mineral, whereas matrix comprises a mixture of discrete fine particles (in arenites, detritus of mixed clay and silt components). Distinction may be difficult, however, because finely crystalline cement with impurities, especially iron oxide stains, may resemble densely packed fine silts or clays with or without impurities. (The average thickness of a thin section is 0.03 mm — finer particles will be superimposed and produce indistinct aggregate optical properties.) Also, pedogenetic/diagenetic processes may produce fine cements (especially carbonates) within the primary matrix, or may transform the primary matrix into interstitial material (such as chert) that resembles primary cement.

Material resembling primary matrix can be produced during weathering and diagenesis by the breakdown of unstable detrital minerals (such as pyriboles, detrital phyllosilicates, and feldspars, particularly the calcic plagioclases). In some cases, intermediate stages in this breakdown can be observed in thin section. Authigenic phyllosilicates may precipitate and/or grow from solution, as well as form by recrystallization of finer detrital phyllosilicates. Clays may be precipitated from colloidal groundwater suspensions, or be translocated by eluviation or illuviation in pedogenesis. In such cases the individual clay particles may pack tightly parallel to framework grain boundaries to form a birefringent *cutan* (often with parallel banding and showing a sinuous band of extinction that moves across the cutan when the stage is rotated UXN); these features may be reorganized during later diagenesis to resemble a primary matrix. Microcrystalline calcite (micrite) resembling the common primary lime-mud matrix of limestones can also be formed during diagenesis by precipitation from solution, or by recrystallization of carbonate grains, or even (rarely) by recrystallization of carbonate cement.

Material resembling precipitated void-filling cement may be produced by diagenetic recrystallization of a primary matrix. For example, lime mud may transform into sparry calcite cement. Chert, a common void-filling cement, also may form during diagenesis from clayey matrix with the concurrent formation of fine chloritic material, illite, or sericite. Relatively large phyllosilicates may form by recrystallization or growth of matrix clays and micas, as well as by primary precipitation. Overgrowths on quartz, feldspar, and carbonate grains may also absorb and replace a primary matrix.

Because of all these problems of distinguishing between primary and secondary matrix and cement, it is probably best to use the soil term *Plasma* in a general, descriptive sense for all interstitial material, and to use the following genetic terms for its inferred character:

Primary cement: mineral matter that has filled original interstitial voids. Use of this term implies the workers' belief that primary cement has not replaced other material. Common attributes include transparent crystals without inclusions; there may be evidence of radial inward growth of the crystals from the edges of surrounding framework grains.

Secondary cement: crystalline mineral matter filling interstitial spaces previously occupied by matrix. The cement may be replacive (expect remnant matrix inclusions) or displacive (expect distorted or compressed surrounding matrix fabrics). Somewhere in the rock unit there should be textural evidence of intermediate stages in the development of secondary cement, but it is sometimes pervasive (as in some dolomitization). It may be indicated if framework grains are "floating" in a cement (that is, if there are exceptionally few or no intergrain contacts).

Primary matrix: fine clastic mineral matter, interstitial to the framework grains, which was deposited in the environment of sedimentation penecontemporaneously with the framework grains. It may be mixed with the framework grains by mechanical means soon after initial sedimentation (for example, by storm- or bio-turbation). It may also be variously recrystallized or even undergo mineralogical changes (for instance, montmorillonites become illite and/or chlorite and/or chert)—in which cases it may be difficult to distinguish it from secondary cement.

Secondary matrix: fine interstitial mineral matter deposited in original void spaces during pedogenesis or diagenesis from colloidal suspensions (such as cutans) or from solution (for example, where unstable framework grains have been extensively altered and the alteration products have reprecipitated). It is difficult to distinguish secondary matrix objectively from primary matrix, particularly where the latter has been recrystallized or mineralogically transformed.

Pseudomatrix: discontinuous interstitial paste formed by the deformation of weak framework grains. Examples are squashed and deformed altered detrital micas, glauconite pellets, and argillaceous rock fragments. Pseudomatrix is commonly evidenced by wisps of material extending from deformed grains into narrowing interstices between framework grains. It is important to record this pseudomatrix as part of the original framework grain population.

INTERPRETATION OF ARENITE COMPOSITION

The composition of detrital sediments reflects the influence of provenance (read Blatt, 1967a), tectonism (Krynine, 1942: Dickinson and Suczek, 1979), and climate (Crook, 1967: Basu, 1976). Composition of the rocks in the source area is of paramount importance (Fig. 76), but the history of earth deformation in the source area broadly dictates the kind of rocks exposed to weathering and erosion (Fig. 78). Intensity of tectonic and climatic factors determine relief; relief and climate determine the balance between rock/mineral decomposition and disintegration, and thus also the products that can be supplied to the depositional basin. Climate-dependent processes also act to modify the products at intermediate sites between the source area and the site of final accumulation. The time factor is important, both in relation to rates of tectonism and of climate-determined processes, insofar as it determines the duration that detritus is subjected to any set of weathering processes. Composition at particular localities will also reflect the effects of selective sorting during transport and other processes acting in the depositional environment (see Davies and Ethridge, 1975). Diagenetic processes acting after final deposition may also modify the compositional attributes of sediments (see separate section). Your textbooks provide general discussions of the controls on sediment composition; also see the selected literature at the end of this chapter, Figs. 76–80, and Tables 14, 15, 16, and 17. Discussion is presented below regarding major arenite clans. Table 18 presents a checklist of essential steps you should take in analyzing any arenite, and Fig. 81 is an example of a laboratory worksheet for description of an arenite.

The most informative components of arenites for provenance and tectonic interpretations are generally rock fragments and heavy minerals (minerals with s.g. greater than 2.85); particular attention should be given to them (see Table 14 and Fig. 75). Unfortunately, the former are absent from many arenites and the latter commonly comprise less than 1% of arenites and thus require concentration after disaggregation of samples to obtain satisfactory and representative assemblages (see later in this chapter). Any chemically unstable or mechanically nonresistant minerals are particularly important for provenance, tectonic, and climatic inter-

pretations. Feldspars may also be very useful (see previous discussion and Fig. 74). The most common component of arenites—quartz—generally requires intensive study and comparison in related sample suites to obtain useful information relevant to provenance/tectonism or the previous history of the assemblage. Recycling of quartz is common and renders interpretation difficult; always evaluate the texture of the quartz grains carefully in conjunction with the textural attributes of the other components when interpreting.

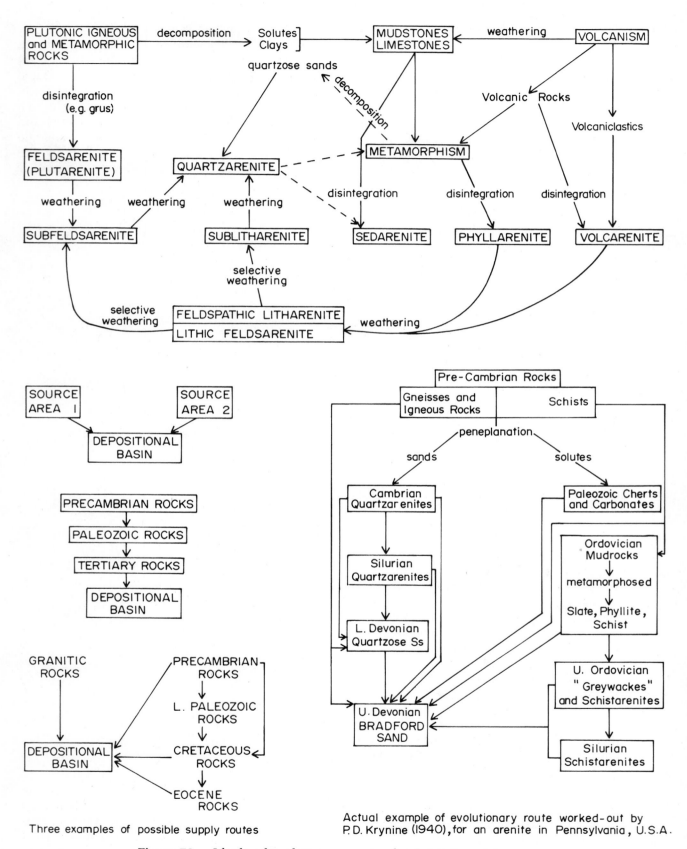

Figure 76. Idealized evolutionary routes for detrital sandstones.

Three examples of possible supply routes

Actual example of evolutionary route worked-out by P.D. Krynine (1940), for an arenite in Pennsylvania, U.S.A.

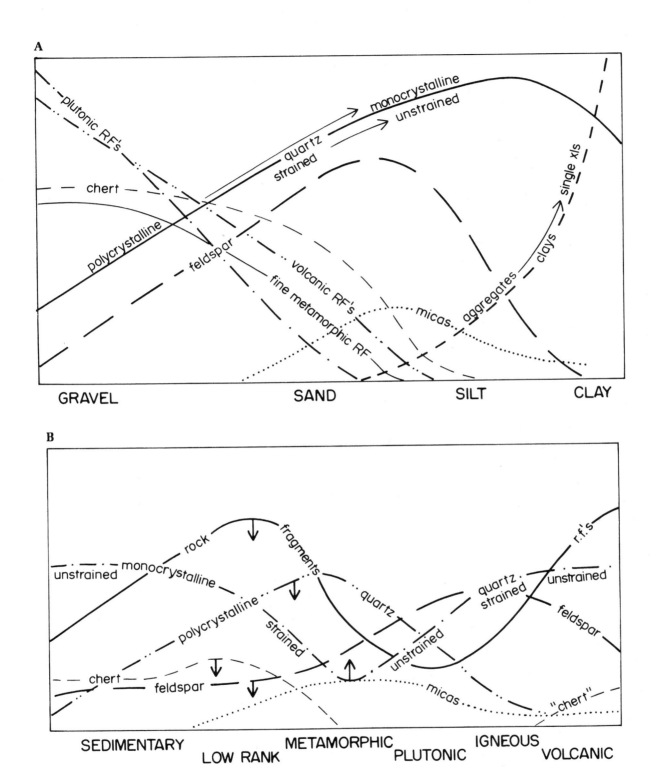

Figure 77. Two controls on abundance of detrital sedimentary components. *A:* Trends of abundance of common detrital components in relation to grain size. Provenance and selective weathering are not the only controls on abundance of components in a sediment! *B:* Expected abundance of sand size detritus as a function of provenance. Curves do not express relative abundance of different components and there are many exceptions. Arrows show whether constituent increases or decreases with weathering.

A

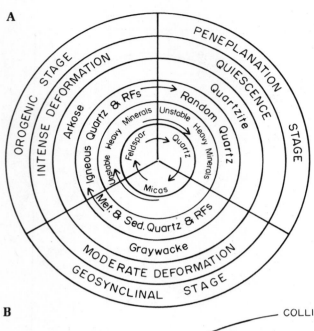

Figure 78. Tectonic cycles and sedimentation. *A:* Krynine's tectonic cycle. *B:* Interpretive megatectonic model—Lewis's 1976 Mark II version. *(A: after Krynine, 1942)*

B

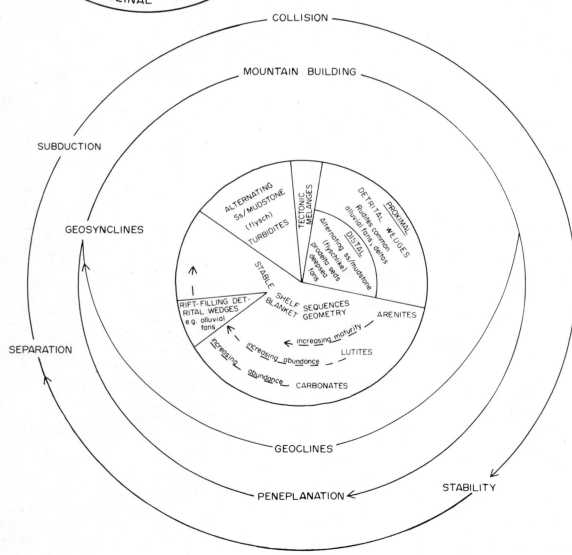

128

Table 15. Empirically Expected Relationships Between Tectonics and Sedimentary Basins

Tectonics of Source Area	Tectonics of Depositional Basin		Results
Orogeny (O) Rapid supply of mineralogically immature sediments		O+R	Thick deposits of mineralogically and texturally immature detrital sediments. Deep and/or shallow and/or emergent paleoenvironments, depending on balance and/or intensity of O to R. Alluvial and submarine fans and deltas with mass flow deposits. Breccias, conglomerates, and arenites proximally, with muds intermixed to dominant distally. Rapid lateral thickness variations.
	Rapid subsidence (R) Rapid burial of sediments, little time for environmental impress, blurred boundaries between paleoenvironments	O+S	Mineralogically immature but may be texturally mature due to reworking. Relatively thin, varied detrital deposits. Most sediment bypassed to more distal depocenters (e.g., via channelized routes as mass flows); typically coarse texturally immature deposits (conglomerates and arenites), with massive homogeneous muds distally.
Epeirogenic uplift (Eustatic fall) (E) Supplies of detritus lacking most unstable components; paleoclimate and time spent in intermediate environments dictate compositional characteristics		E+R	(rare) Texturally immature (due to rapid final deposition) to mature (due to previous history); mineralogical maturity depends on previous history. High proportion of nondetrital sediment, e.g., bioclastic limestones; reefs along hingeline. Arenites and mudstones commonly in deltas and/or submarine fans.
		E+S	Texturally mature detrital sediments of distal alluvial to shallow marine paleoenvironments. Cyclothems. Arenites proximally to mudstones distally.
	Slow subsidence (S) Slow burial of sediments, environmental processes with strong effect on sediment characteristics; sharp boundaries between paleoenvironments, which cover broad areas	Q+R	Proximal texturally and mineralogically mature to supermature arenites. Evaporites where arid and restricted circulation; black shales where temperate and restricted circulation. Hingeline reefs, some mass flow deposits seaward of hingeline. *Starved basins* with deepwater carbonate and siliceous oozes.
Quiescence (Q) Minor supplies of mineralogically mature detritus		Q+S	Texturally and mineralogically supermature arenites. Common siltstones and claystones. Abundant carbonates, either bioclastic or lime mudstones. Shallow to moderate-depth marine deposits (shelf-type).

The tectonics of the source area tend to dictate mineralogical maturity, whereas the tectonics of the depositional basin tend to dictate textural maturity.

Note: The term "basin" does not imply a circular or trough configuration, nor necessarily a bathymetric/topographic depression; "depocenter" is probably a better word for use in technical writing.

Rifted Continental Margin

(a) Early phase

(b) After separation

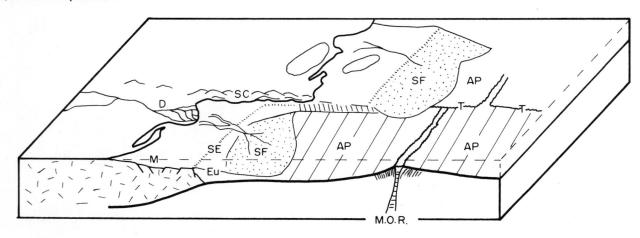

M   = Miogeocline

SE = Shelf edge (in modern settings; some
      question if "shelves" always existed as such

Eu = Eugeocline

AP = Abyssal Plain

T   = Transform

D   = Delta

SF = Submarine Fan

SC = Steep Coast, narrow shelf, etc

M.O.R. = Mid Ocean rise/ridge

Figure 79.   Sedimentary basins in plate tectonic settings. *(Sketches by J. D. Bradshaw, University of Canterbury)*

CONVERGENT MARGIN SETTING

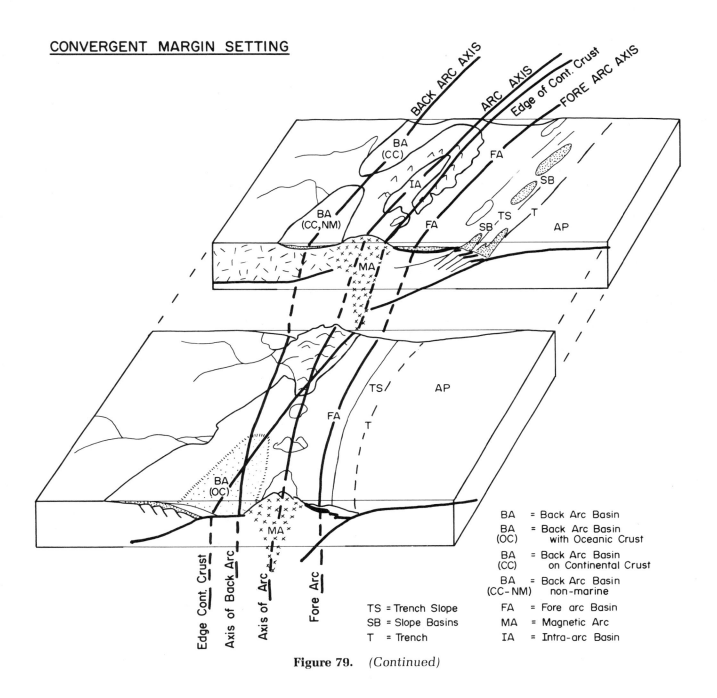

BACK ARC AXIS

ARC AXIS

Edge of Cont. Crust

FORE ARC AXIS

BA
(CC)

FA

IA

SB

BA
(CC,NM)

FA

SB

TS

T

AP

MA

TS

AP

FA

T

BA
(OC)

MA

Edge Cont. Crust

Axis of Back Arc

Axis of Arc

Fore Arc

TS = Trench Slope
SB = Slope Basins
T = Trench

BA = Back Arc Basin

BA
(OC) = Back Arc Basin
with Oceanic Crust

BA
(CC) = Back Arc Basin
on Continental Crust

BA
(CC-NM) = Back Arc Basin
non-marine

FA = Fore arc Basin
MA = Magnetic Arc
IA = Intra-arc Basin

Figure 79.   *(Continued)*

Table 16. A Plate-Tectonic Classification of Sedimentary Basins

Basin Setting	Tectonism	Sedimentation
Rifted Continental Margins	Pre-rift arching	Thin alluvial fans on flanks of arch
	Rift valley grabens, and fault-angle depressions ↓ Proto-ocean ↓ Narrow ocean ↓ Open ocean, miogeosyncline becomes separate miogeoclines and eugeoclines	Thick alluvial/deltaic wedges, lacustrine deposits (tuffs, evaporites, diatomites). Coarse marginal deposits interfinger with central finer sediments. Some volcaniclastics. Climate important. Oceanic influence grows as rift opens. May have euxinic or evaporitic "starved" marine basins. However, generally sediments from continent provide detritus for stable to unstable "shelf" miogeoclinal sequences, tending coarse to fine both laterally and vertically, with carbonates often dominant distally or when relief is low. Eugeoclines of continent-derived detritus plus some volcanics supplied by seafloor spreading; submarine fans.
	Failed rift valley	Commonly thick section with early volcanic phase (often rhyolitic), later arenites and/or carbonates. Finally filled by detritus from continent.
Convergent Margin Settings	Trench basin, in subduction zone	Melange on inner margin, basin fill of mixed pelagic/turbidite sequences "scraped off" the consumed seafloor. Submarine fans. Parts of ophiolites.
	Fore-arc basin, between trench and magmatic arc	Thick fluvial to deltaic to neritic to deep marine sequences, variable due to rate of compression and variable history. Eugeoclinal.
	Intra-arc basin, between continent and arc or two island arcs	Eugeosynclinal volcaniclastic sequences, with continental detritus in one case.
	Back-arc basins— Inter-arc, on oceanic crust behind arc	Volcaniclastic eugeoclinal turbidite-rich sequences, minor pelagic sediments.
	Retro-arc, on continental crust behind arc	Alluvial-deltaic-shallow marine sequences to considerable thicknesses, derived from fold/thrust orogen.
	Suture belt, between crustal blocks; "peripheral basins" possible at junction	Complexly deformed sequences similar to those in retro-arc basins.
Intra-continental	Many may reflect crustal attenuation during early stages of rifting that subsequently failed. Others not easily related to plate-tectonic theory. Commonly surrounded by stable continent. Alluvial to marine sequences; arenites-mudstones-common carbonates.	
Oceanic	Rise crest, at mid-oceanic rift Rise flank Deep basin	Volcanic turbidites and pelagic carbonates; siliceous oozes

Source: after Dickinson, 1974

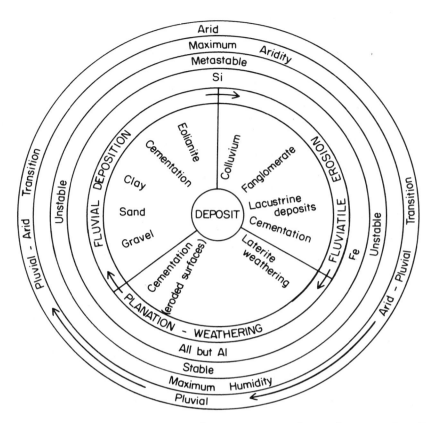

Figure 80. Idealized climatic cycle and continental sedimentary products.

Major indicators of paleoclimate in sediments: *Fossils,* especially flora (e.g., palynology), also stenothermal fauna and fauna with special paleoecological requirements. Main problems: redeposition; uncertainties prior to Tertiary. *Oxygen isotopes,* O^{18}/O^{16} ratio in carbonates. Problems: some organisms do not precipitate $CaCO_3$ in equilibrium with the water; diagenetic changes in the ratio. *Trace elements,* e.g., Sr/Ca and Mg/Ca ratios in carbonates. Problems: original shell mineralogy influences ratios; diagenetic changes in the ratios. *Clay mineralogy,* see later section. Problems: redeposition; diagenetic transformation. *Sedimentary iron* compounds, see later section. Problems: diagenetic and epigenetic transformations. *Type and extent of detrital mineral alteration,* imprecise and difficult to apply. See charts of relative mineral stabilities. If provenance is known, *absence* of minerals in the derived sediments may be indicative. Problems: *time* of exposure to chemical weathering vs intensity of weathering conditions; diagenetic alterations. *Miscellany:* evaporites (particularly occurrence of those more soluble than Ca-sulfates); glacial deposits (problems of distinction from plant-rafted dropstones, sediment gravity-flow deposits). *(Diagram after M. J. Carr, 1966, B.Sc. thesis, Australian National University; Chavaillon, 1964; see Alimen, 1965)*

Table 17. Mineral Stability Relationships under Sedimentary Conditions

Goldich Stability Series (generalization for common rock-forming minerals, effectively the inverted Bowen's Reaction Series without reactions)

	Quartz	
	Muscovite	
	K-feldspar	↑ Stability
Biotite	Na-plagioclase	
Hornblende	Na-Ca plagioclase	
Augite	Ca-Na plagioclase	
Olivine	Ca plagioclase	

Relative Chemical Stability of Some Minerals (see Brewer, 1976, for sources and discussion)

Pettijohn (1957, 1975)		Weyl (1952)	Graham (in Jackson and Sherman, 1953)	Marel
anatase muscovite rutile	} more common in ancient rocks than modern sediments—some are authigenic	rutile, zircon tourmaline, sphene, magnetite kyanite, andalusite sillimanite epidote, garnet augite, hornblende olivine	quartz, muscovite, rutile, zircon, tourmaline, ilmenite, andal- usite, kyanite, sphene, magnetite staurolite biotite, epidote, garnet, augite, hornblende apatite, olivine	tourmaline rutile staurolite zircon garnet muscovite epidote amphibole augite biotite hypersthene olivine
zircon tourmaline monazite garnet biotite apatite ilmenite magnetite staurolite kyanite epidote hornblende andalusite topaz sphene zoisite augite sillimanite hypersthene diopside actinolite olivine				

Table 17. (Continued)

Fieldes and Swindale (1954)	Smithson (1941)	Dryden and Dryden 1946	Clay-size minerals (after Jackson and Sherman, 1953)
quartz	zircon, rutile, tourmaline, apatite	zircon	anatase, zircon, rutile, ilmenite, leucoxene, corundum, etc.
feldspars, acid volcanic glass	monazite	tourmaline sillimanite monazite	hematite, geothite, limonite, etc.
muscovite, biotite	garnet, staurolite, kyanite	chloritoid kyanite	gibbsite, boehemite, allophane, etc.
zeolites, basic volcanic glass	mafic minerals	hornblende staurolite garnet	kaolinite, halloysite, etc.
augite, hornblende, hypersthene, olivine		hypersthene	montmorillonite, beidellite, saponite, etc.
			mixed 2:1 layer clays & vermiculite
			muscovite, sericite, illite
			quartz, cristobalite, etc.
			albite, anorthite, stilbite, microcline, orthoclase, etc.
			biotite, glauconite, Mg-chlorite, antigorite, nontronite, etc.
			olivine, hornblende, pyroxenes, diopside, etc.
			calcite, dolomite, aragonite, apatite, etc.
			gypsum, halite, Na-nitrate, ammonium chloride, etc.

Relative Resistance to Abrasion of Some Minerals (see Pettijohn, 1975, for sources and discussion)

after Friese (1951)		after Thiel (1945)	
tourmaline	epidote	quartz	rutile
pyrite	olivine	tourmaline	hypersthene
staurolite	apatite	microcline	apatite
augite	kyanite	staurolite	augite
topaz	andalusite	sphene	hematite
magnetite	orthoclase	garnet	kyanite
garnet	monazite	epidote	fluorite
ilmenite	hematite	zircon	siderite
		hornblende	barite

Note: Significant differences shown between the relative stabilities of some minerals probably reflect different chemical conditions in the settings studied. Hence, depending on the pedogenetic/diagenetic conditions, relative positions of some minerals change.

Spl. No: Formation: Location:

Petrologist: Date:

<u>HAND SPECIMEN</u> Colour: Fresh-   Weathered-

Induration: Structures:

Grain Size: Max.
             Min.   Roundness
             Mode(s)   Sphericity

Sorting:
Other Features:

Classification:

<u>THIN SECTION</u> Apparent Grain Size:Max.   Roundness
                                     Min.
                                     Mode(s)   Sorting

COMPOSITION   PERCENT   RECALCULATED PERCENT FOR FOLK <u>et al.</u>
                                                        CLASSIFICATION

Quartz   (primary   (secondary
Feldspar   triangle)   triangle)
Rock Fragments
Micas   Q
H & O Minerals
Others   F

                                         R
                                              ‾‾‾‾‾‾‾   ‾‾‾‾‾‾‾
                                              100%   100%

Matrix   Predominent F grains
Cement   Predominent R grains

                                                                        TOP

              ‾‾‾‾‾‾‾
              100%

H & O Varieties:

Quartz Varieties:

Diagenetic Modifications:

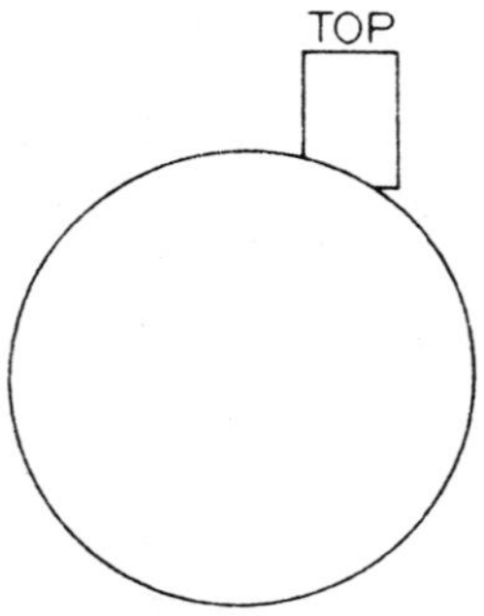

Other Features: X:
                                         Diameter:
                                         Illustrating:

FINAL ROCK NAME:(Colour)   (Induration)   (Structures)
(Sorting Term)   (Size Term)   : (Cement)   (Prominent nondetrital
grain type)   (Prominent detrital grain type, not Q, F, R)
(named arenite)

Figure 81. Example of laboratory worksheet for petrographic analysis
of arenites.

Table 18. Description/Discussion Format for an Arenite

Formal description/discussion should follow a systematic pattern, combining field/hand specimen/thin section/refined analytical determinations.

Final rock name, i.e., classification

Structures:
- description at all scales, field and laboratory
- paleocurrent and/or way-up indications
- vertical and lateral associations

Textures:
- modal size; range of sizes and their proportions
- roundness—any variation with size or composition?
- sorting of sand sizes; proportion and character of matrix
- textural maturity—any kind of "inversion"?
- fabric—anything notable?
- sphericity—discussed only if unusual
- porosity—present and primary (now cement); see Halley (1978)

Composition:
- percentages of mineral/RF types; relationship of each to grain size
- quartz characteristics and proportions (mono- vs. polycrystalline, strained vs. unstrained)
- feldspars—unless a detailed study, discriminate twinned vs. untwinned vs. microcline vs. altered (original variety unrecognisable); degree of alteration of recognisable varieties
- rock fragments—characteristics of each variety; total percentage or relative proportions or each; size and shape
- heavy minerals—varieties and proportions, roundness
- opaque minerals—varieties and proportions, roundness
- perigenic clasts—polymineralic (mud galls, etc.); monomineralic (glauconite, etc.)
- matrix—identification to appropriate level

Diagenesis:
- cement character and proportion; localization
- matrix-modifications from initial character
- replacement and alterations of each variety of mineral component
- overgrowths on quartz/feldspar; abundance
- pressure-solution effects?
- authigenic minerals—crystal form and textural character relative to other components
- recrystallization or crystal growth features (such as micas growing from clays)

Synthesis/Interpretations:
- structures—regarding energy levels and processes during deposition
- textures—regarding depositional processes, previous history of grains (may require synthesis with composition); regarding diagenetic modification
- composition—regarding provenance, paleoclimate, paleotectonism (will require synthesis with textures); regarding depositional environment (physical/chemical character of perigenic fraction); regarding diagenetic environment (physical and chemical implications of sequential modifications to minerals, including cements)

Quartzarenites

Quartz is the most stable of the common rock-forming minerals, but it does not dominate in any volumetrically significant primary source rock. Hence quartz-rich sediments are generally produced as a result of preferential destruction of other minerals during the sedimentary cycle. Chemical weathering eliminates the other relatively unstable minerals by various reactions (such as hydration, hydrolysis, carbonation, and cation exchange): products are solutes and secondary minerals (such as clay minerals and iron oxides) that are finer than sand size. Physical weathering breaks the less-resistant minerals into particles finer than sand size (most of the common rock-forming minerals have a better-defined cleavage and are more brittle than quartz) and in the process renders the particles more subject to chemical weathering (by increasing the surface area per unit volume). Quartzarenites therefore reflect intensive chemical weathering of source rocks or the grain assemblage after derivation from the source rocks, generally but not necessarily accompanied by a prolonged history of physical weathering. Very prolonged exposure to lower-intensity chemical weathering is also possible. Grain roundness helps distinguish between these alternatives, since rounding mainly results from mechanical abrasion. (Some rounding may be due to chemical solution, in which case the grains will be embayed and there should be a similar effect on all grain sizes present, whereas if rounding is due to abrasion, finer grains should be less rounded.) In either case, tectonic quiescence is generally indicated. However, most quartzarenites, particularly those with rounded grains, probably reflect recycling from an immediate source rock of older quartz-bearing sediment. If such a source is inferred, tectonic activity (rejuvenation) is likely, and less intense chemical and physical weathering is required during the final cycle. Well-rounded quartz requires a truly extensive history of abrasion, rarely achieved in one sedimentary cycle, however long. Localized exceptions may exist in the case of quartz-rich sands subjected to prolonged eolian or beach processes.

Any rock fragments (such as those in interlayered conglomeratic beds) or heavy minerals present in the quartzarenite are important indicators of the sediment history. Only the chemically ultrastable grains (chert, zircon, tourmaline, rutile, hematite, magnetite, and ilmenite) are expected in the case of sediments that have suffered the most prolonged history of weathering. The *ZTR index*—a measure of the percentage of the ultrastable nonopaque heavy mineral suite of zircon, tourmaline, and rutile —commonly is calculated to indicate the degree of mineralogical maturity of the sediment (see Hubert, 1962). Survival of other minerals that are resistant to physical abrasion but less resistant to chemical weathering suggests prolonged exposure or recycling under conditions of relatively low-intensity chemical weathering. For example, the presence of rounded feldspars suggests an arid paleoclimate. The relative abundance of polycrystalline and strained quartz may be indicative of abrasion history, because they are preferentially destroyed relative to unstrained monocrystals; however, their original abundance and character are dictated by provenance. Plots of "stable" versus "unstable" polycrystalline quartz grains (see Young, 1976; Mack, 1981) or of undulose (extinction >5° stage rotation) versus nonundulose (extinction <5°) versus polycrystalline (2–3 vs >3 crystal units per grain) may in fact prove useful in provenance determinations (see Basu et al., 1975 and Fig. 82). (*Note:* When studying the relative proportions of these quartz varieties, you must make comparisons within the same size fractions. In addition, when provenance is uncertain, interpretations from such studies usually must be based on comparisons of the relative proportions of the varieties in arenites from different stratigraphic levels.)

Most quartzarenites form under stable, quiescent tectonic situations on, or on the edges of, cratons. A sheet geometry, reflecting nearshore deposits of marine trangression during epeirogenic subsidence (or occasionally, regression during tectonic rejuvenation), is the most common shape of quartzarenite deposit; some occur as "shoestring" beach or bar deposits in more active tectonic regimes. Most of these deposits are texturally mature or supermature, but textural "inversions" (such as bimodal grain roundness, or bimodal size distributions) are common. The amount of clayey matrix, in particular, varies within this mineralogically mature suite. Diagenetic destruction of accessory minerals and development of grain overgrowths are common because a high porosity and permeability facilitate fluid circulation. In some cases, original subfeldsarenites are converted to quartzarenites by the diagenetic (or epigenetic) destruction of feldspars; indication of this change may be an irregular distribution and size of voids or the presence of disseminated clay patches (often lost during thin-section preparation). Cements of quartzarenites are most commonly silica (usually as grain overgrowths) or carbonates. Disseminated specks of hematite may

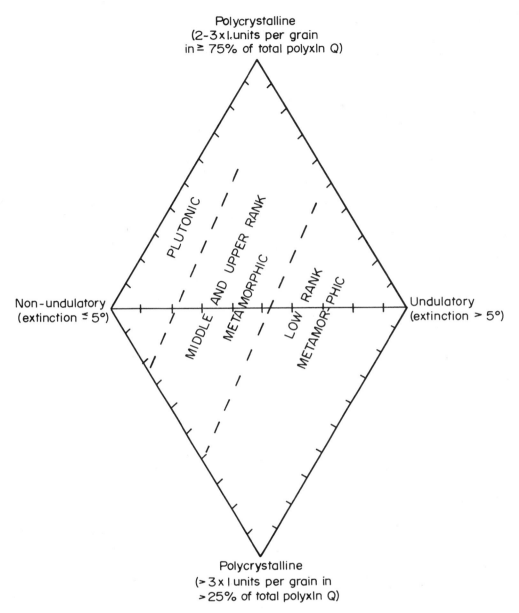

Polycrystalline
(2-3 x l.units per grain
in ≥ 75% of total polyxln Q)

PLUTONIC

MIDDLE AND UPPER RANK

METAMORPHIC

LOW RANK METAMORPHIC

Non-undulatory
(extinction ≤ 5°)

Undulatory
(extinction > 5°)

Polycrystalline
(> 3 x l units per grain in
> 25% of total polyxln Q)

Figure 82. Graph for plotting quartz varieties in arenites, showing fields characteristic of particular source rocks. Consult Basu et al., 1975, for limitations in the application of this method. *(After Basu et al., 1975)*

obscure the nature of clayey matrix or carbonate cement and impart a distinctive red color to the rock; this hematite may form in the depositional environment, whereupon the specks are commonly concentrated on the rims of the detrital grains, or from diagenetic or epigenetic alteration of iron-bearing minerals (relocation in reducing or acidic solutions may disseminate ferrous iron prior to oxidation and precipitation).

Feldsarenites

Feldspar is the commonest mineral in most primary source rocks. Whenever chemical weathering is restricted, either because of a rigorous climate or because too little time is available, feldspars should also be abundant in the sand-size fraction. Apart from chemical weathering, sedimentary processes only reduce the relative abundance of feldspars in arenites by physically breaking them into sizes finer than sand. All gradations from feldsarenites to subfeldsarenites are common, depending on the size fractions studied, proximity to the source, and the geological history of the sediment.

Tectonic feldsarenites: produced when the rate of sediment supply is too rapid for chemical destruction of the feldspars. Characteristically, tectonism will have created a relatively high relief, such that physical processes of erosion dominate and grains are rapidly moved from the source rocks into the site of deposition and rapidly buried there. Block faulting, resulting in the exposure of plutonic rocks, is the typical cause. Conglomerates (and breccias) are characteristically associated, and the rock fragments therein provide good evidence of provenance; however, coarse deposits may be sparse in the more distal or stratigraphically highest parts of the succession. Muds are common as a matrix to the sandstones, and as silty interbeds; commonly there is a regular (or irregular) repetition of sandstone-mudstone units. The overall geometric pattern of the feldsarenites and associated facies is characteristically wedge-shaped (as in alluvial fans). Individual beds are typically discontinuous and show channel structures; the more continuous ones commonly result from lateral migration of channels and are thus diachronous. Nonmarine deposits are generally abundant and are commonly red because of deposition and diagenesis under oxidizing conditions, but the more distal deposits commonly show drab colors because these environments are marine or have a consistently high groundwater table level, which inhibits oxidation. Subsequent to the middle Paleozoic, temperate-region terrestrial deposits commonly contain abundant plant remains, which may establish and maintain reducing conditions. Grains are typically angular and many of the deposits are texturally immature. Rock fragments (especially polycrystalline feldspars or feldspars plus quartz or mica) tend to be common in the coarser sand grades, and chemically unstable minerals (such as biotite or the pyriboles) are often present. Feldspar grains are typically fresh, but some may be partly weathered (particularly the calcium-plagioclases, but see Todd, 1968); in many cases both fresh and partly weathered feldspars of the same species are present, because different grains have been exposed to different degrees of chemical weathering.

Climatic feldsarenites: produced when climatic conditions are sufficiently rigorous (arid or cold) that chemical weathering is inhibited. In constrast to the previous situation, quiescence or slow epeir-

ogenic movements prevail and the grains are subject to extensive physical reworking before they are finally buried; hence grains are commonly rounded. Conglomerates are sparse, relatively fine-grained, and consist of stable minerals and rock fragments (such as chert, polycrystalline quartz, and aphanitic acidic volcanics). The overall geometry is sheetlike, and cross-bedding is typical; channeling may be abundant. Whereas final accumulation may occur under transgressive (or regressive) marine conditions, the environments in which these feldsarenites acquire many of their characteristics were probably analogous to deserts like those of central Australia today. However, prior to the development of extensive terrestrial vegetation (which acts to bind sediment and retain water, as well as to contribute powerful chemical weathering agents via its root action and decay products), it is probable that suitable conditions existed in more temperate and humid climates than today. Certainly such deposits are much more common in Precambrian and Lower Paleozoic sequences than in younger ones (for example, see Lewis, 1971). Deposits are commonly red, but they may be white, in which case they can be commonly mistaken for quartzarenite in the field. Most are texturally mature or supermature; bimodal size distributions are common. Muds are generally rare and well segregated (probably due to deflation), whereas gravels may commonly interfinger with the sands. Minerals chemically more unstable than the alkaline feldspars are generally lacking (although chemical weathering processes are relatively restricted, the sediments have been exposed to them long enough for widespread destruction of the unstable components to have occurred). The ultrastable minerals (zircon, tourmaline, rutile) often dominate the heavy mineral suites, but less stable minerals (such as garnet and apatite) should also be present. Microcline is typically fresh, orthoclase fresh to weathered, and most plagioclase altered beyond the level of species identification. Authigenic mineral developments (such as quartz and feldspar overgrowths) are generally abundant, and diagenetic destruction of chemically unstable minerals may be extensive (as with quartzarenites, most deposits are highly porous and permeable).

Recycled feldsarenites: produced when tectonism uplifts poorly indurated feldspathic sediments and/or volcanics, either in a rigorous climatic setting where chemical weathering is inhibited or where transport and deposition are very rapid. The tectonic setting is commonly dominated by fold deformation, which creates highs that are continuously eroded during uplift and depositional troughs that are continually sinking to receive the detritus; however, early deposits from fault-block movements, or deposits from rapid epeirogenic uplift, may fall into this category. Because of the intermediate character of the tectonic setting (neither broad, slow epeirogeny nor extensive vertical uplift to expose plutonic basement), the geometric characteristics of the deposits are intermediate between the tectonic and climatic feldsarenites. Deltaic and submarine fan deposition of flyschlike successions with thick alternating sandstones and mudstones are most likely. Textural characteristics are varied, depending on the previous sedimentary history of the grains and the depositional locale, but grains are typically subground. Conglomerates, with dominant sedimentary and/or volcanic rock fragments, may be common or sparse. Compositional criteria are most diagnostic. Characteristically, there will be sedimentary (and volcanic) rock fragments present—all gradations to, and associations with, sublitharenites, lithic feldsarenites, and sedlitharenites are likely. Feldspars are typically altered, although some K-feldspars may appear fresh. Distinction between altered feldspars and sedimentary rock fragments is often difficult, as is distinction between these and any muddy matrix that may be present.

Igneous feldsarenites: produced essentially in association with intermediate to basic igneous activity, and hence mainly near island arcs or plate boundaries. These sediments may be derived by winnowing from extrusive volcanic deposits, in which case they show gradation to, and association with, volcarenites. Or they may be derived from granodiorites and more basic intrusives, in which case plutonic rock fragments (such as polycrystalline feldspars) are generally present in the coarser grain sizes. Distribution and characteristics of these deposits are not well known—they are rare away from sequences that have been extensively deformed (and most have probably been metamorphosed out of the sedimentary class). Where recognized to date, these rocks are mainly associated with volcarenites and recycled feldsarenites—hence their history appears intimately interrelated. Compositionally, they are characterized by a dominance of calcic plagioclases, and because of t e ease with which these minerals are destroyed by chemical weathering, rapid transport and burial are necessary for preservation. Grains are thus angular, and texturally immature deposits are most common; however, well-sorted igneous feldsarenites are known (for example in the Mesozoic rocks of New Zealand). Alteration of feldspars to various clay minerals, resulting from either weathering or diagenesis, sometimes renders obscure their differentiation from clayey matrix or fine-grained rock fragments.

Litharenites

Most litharenites are deposited during times of orogenic activity, the main exceptions being volcarenites. Physical weathering processes are as effective as chemical weathering in destroying rock fragments (and are more effective in the case of most sedimentary clasts). Abrasion stresses the internal planes of weakness between crystal individuals, and the rock fragments progressively disintegrate into smaller grains until monocrystalline mineral fragments alone remain (or until the particles are finer than sand). Fig. 75 and Table 14 illustrate how textures of the coarsely crystalline parent rocks are lost by the time grains are in the +2 to +3$\varnothing$ range. Hence, coarse-grained litharenites commonly are associated with fine-grained sublitharenites, and fine-grained sublitharenites commonly result from the same geological history as coarse litharenites. In some litharenites (such as in many graywackes), differentiation of fine-grained rock fragments from clayey matrix is difficult, a problem that is increased by the creation of clayey matrix by the chemical diagenetic destruction of unstable accessory minerals.

Volcarenites warrant little discussion—their implications are obvious. Because of the fine interlocking fabric of the parent rock, clasts tend to survive considerable abrasion and retain recognizable textures into the fine sand grades. Chemical weathering is effective in destroying the more basic volcaniclasts, and it commonly alters the initial mineral components such that the clasts become difficult to distinguish from other finely textured clasts (such as mudstones, chert, and phyllitic metamorphics—see Fig. 73). Diagenetic modifications (such as silicification and alteration analogous to chemical weathering) also may render discrimination difficult. Mineralized vesicles, devitrified glass texture, and remnant lathlike feldspars are the main recognition criteria. Volcanic clasts may dominate in an arenite or constitute only a fraction of it, depending both on proximity to extrusive centers and on geological history; wherever they are present, much larger volumes of volcanic ash (now indistinguishable from sedimentary muds) are likely to be associated (such as in lateral facies). When volcanic clasts dominate, there is the possibility that the deposit represents a direct airfall volcanic tuff; sedimentary structures and textures (such as grain roundness and sorting) are criteria in these cases, and all gradations from direct airfall to exclusively sedimentary deposits exist.

Sedlitharenites are characteristic of the early phases of block faulting and of fold deformation; in the former case these are commonly overlain by plutarenites and associated with conglomerates, whereas in the latter case conglomerates may be few and overlying arenites may have progressively greater proportions of low-grade metamorphic detritus (which is often difficult to distinguish in thin section from sedimentary rock fragments). Epeirogenic rejuvenation may also produce some sedlitharenites, but erosion in such cases is often slow enough to disarticulate the rocks into their original monocrystalline particles.

Sandstone-litharenites are rare because clasts commonly break down into monocrystalline sand grains; however, clasts of granule and very coarse sand size are common in a few deposits. When the source rock is a silica-cemented quartzarenite, clasts of all sizes may be common, but problems arise in distinguishing such clasts from plutonic polycrystalline quartz and metaquartzite fragments. When the sandstone clasts have a muddy matrix, look for evidence to prove that they are detrital rather than perigenic.

Mudrock-litharenites are common in some sequences; mudrock clasts are not much affected by chemical weathering (there may be oxidation rims) and may be very tough (as when they are silicified). The main problem in thin section becomes one of distinguishing such clasts from altered volcanics and low-grade metamorphics (see Fig. 73), or from any muddy matrix that may be present in the litharenite (hold the thin section of coarse varieties up to a light source to distinguish boundaries). Evidence must also be sought to distinguish detrital from perigenic mudstone clasts—the latter are common in many arenites, although mainly in the gravel sizes.

Calclitharenites are produced only where erosion, transportation, and deposition (into a marine basin) are very rapid; hence they are characteristic of block-fault tectonic settings. Carbonates are very unstable in the terrestrial weathering environment (rainwater and most groundwaters are acidic), and the carbonate minerals are relatively soft (hardness of 3-4). Calclithrudites are generally intimately associated with these rocks and often dominate volumetrically (they may also be produced—albeit rarely—in the submarine environment, whereas submarine production of calclitharenites is most unlikely). A major problem is to distinguish detrital from perigenic carbonate clasts (intraclasts)—most perigenic clasts have a lime-mud matrix, but recrystallization may occur during diagenesis.

Chert-arenites are uncommon, but chert sublitharenites are not. Chert is chemically and physically very stable and resistant, thus it survives into even the finest sand sizes. It is not common as a rock type (hence chert-arenites are uncommon), but it does occur abundantly in some local successions (mostly associated with limestones, but occasionally as thick *novaculites*). Chert is a diagenetic product representing the replacement of mud, carbonate, or other material; the necessary silica is most commonly remobilized from organic hard parts (diatoms, radiolaria, sponge spicules, and the like). Some silica may be supplied by groundwaters from outside (as in silicified soils, or *silcretes*); some may come from the alteration of unstable silicate minerals or the pressure-solution of grains within the same rock succession. In the rock being studied, such diagenetic silicification of muddy matrix or volcanic clasts can render distinction from detrital sedimentary chert difficult. Chert clasts are almost as resistant as quartz grains, hence they may be an indicator of sediment recycling—as long as you are sure they are sedimentary (see Fig. 73).

Phyllarenites are produced when low-grade metamorphic rocks are strongly uplifted and rapidly supplied to the site of deposition. Because of the micaceous foliation, clasts are relatively easily destroyed by physical abrasion; conglomerates of these clasts are uncommon. The common chloritic composition of many such clasts permits relatively easy chemical weathering to clay minerals, and both weathering and diagenetic alteration can make the clasts very difficult to distinguish in thin section from mudstone or volcanic clasts. Phyllarenites may be produced in the penultimate stages of complex fold-dominated orogeny, in which case they will stratigraphically overlie recycled sediments (such as recycled feldsarenites and sedlitharenites) and underlie plutarenites. Or they may be produced by fault-block rejuvenation of regionally metamorphosed sequences.

Folk (1974) distinguished miogeosynclinal (miogeoclinal), eugeosynclinal (eugeoclinal), and rejuvenation phyllarenites. Miogeoclinal types form along the continental margin, are derived from horizontally compressed, uplifted belts of low-rank metamorphic rocks, and are deposited in rapidly prograding coalescent alluvial and deltaic or shallow marine sequences. Rapid lateral changes in sediment characteristics are expected, because the clasts are easily abraded; both composition and texture will become more mature progressively as more energy is expended on the sediments. Most sands will be fine-grained because of the ease of breakage, and texturally supermature deposits may occur in high-energy environments such as beaches. Eugeoclinal types accumulate in rapidly subsiding troughs next to subduction zones and have abundant supplies of volcanic detritus; a relatively deepwater environment is suggested, which is commonly dominated by turbidite successions. Higher-grade metamorphic rocks contribute more than in the case of miogeoclinal types, and compositional variation may range from plutarenites (schist- or gneiss-arenites) through phyllarenites to volcarenites. Rejuvenation phyllarenites are produced from the uplift of old metamorphic rocks—that is, metamorphism does not accompany uplift. Block faulting is the common cause of rejuvenation in these cases. Most deposits will probably be subphyllarenites, and they may be associated with (older) sedlitharenites and (younger) feldsarenites.

Plutarenites are produced when crystalline plutonic rocks are rapidly uplifted and eroded. In situations where chemical weathering dominates, these source rocks are disarticulated into monomineralic clasts of sand or finer sizes (and/or altered into secondary mineral assemblages). Hence plutarenites are generally coarse grained and are characteristic of rapid supply from block-faulted areas or areas in the final phases of complex orogeny. In the former case, they commonly occur stratigraphically above thin sedlitharenites or quartzarenites (reflecting progressive unroofing of the "basement"), but in old (for example, peneplaned) cratonic regions there may not be a sedimentary cover and the plutarenites may be the first deposits. In the latter case, they commonly overlie a thick succession of recycled sedimentary deposits (recycled feldsarenites and sedlitharenites) and phyllarenites. Or they may extend in wedge-shaped deposits over relatively distal mudrock successions, since they reflect prograding paraclysmal or post-paraclysmal orogenic deposits (molasse). All gradations to feldsarenite and sublitharenite are common, and associations with rudites composed of plutonic clasts can also be expected. Plutarenites are generally texturally immature, although some well-sorted deposits do occur.

LABORATORY TECHNIQUES FOR ARENITE ANALYSIS

Preparing Thin Sections

THIN-SECTIONING INDURATED INSOLUBLE ROCKS

1. Cut the rock with a diamond saw to produce a 3-mm-thick slice with parallel sides.

2. Grind the slice on one side, on a coarse grinding lap using 180-grade and then 320-grade carborundum powder with water as a lubricant to remove diamond saw marks. Rinse thoroughly after each grinding operation.

3. Continue grinding on a glass plate using 400-grade and, after rinsing, 600-grade carborundum with water. Use a rotary action until you produce a finely finished flat surface.

4. Scrub with running water to remove waste carborundum and then dry the slice on a hotplate.

5. Mount the slice on a clean microscope slide with "Lakeside 70" thermoplastic cement, using a hotplate set at 100°C. Apply the cement to the polished face of the hot rock slice and press the slide down until all excess cement and any bubbles are squeezed out.

6. Grind the upper surface of the rock slice on a coarse grinding lap using 180 and then 320-grade carborundum with water. Continue until you can see light through the rock.

7. Grind the surface on a glass plate using 400 and then 600-grade carborundum with water. Use a rotary action and make sure that you keep the surface parallel to the surface of the microscope slide. Check the thickness regularly, using a polarizing microscope, until you obtain a final thickness of 0.03 mm. Judge the correct thickness by examining the birefringence—quartz, for example, will show a pale, first-order gray color when it is 0.03 mm thick.

8. Wash the thin slice and then dry it thoroughly. Coat it with cellulose acetate to prevent it from breaking up, and trim it to a convenient size with a razor blade.

9. Heat the thin rock slice on a hotplate set at 85°C, then transfer it from the old slide to a new microscope slide with canada balsam on it. The balsam should be heated initially for about 3 min,

or until a bead, picked up on the points of tweezers, forms a brittle thread that snaps on cooling when the tweezers are opened. Place more canada balsam on the upper surface of the thin slice, and then cover it with a coverslip. Then clean the completed thin section with alcohol or acetone and label it. (Canada balsam is very expensive and some laboratories use other mounting media; if another medium is used, hotplate temperatures may differ and you should discover its refractive index before analysis with the petrographic microscope.)

THIN-SECTIONING FRIABLE, WATER-SOLUBLE, AND SWELLING ROCKS

The basic technique is the same as that described above, but various additional treatments are necessary. For example, friable rocks must be impregnated with a cement to consolidate them; water-soluble rocks and those with swelling components require kerosene or glycol to be used as a lubricant instead of water. See Allman and Lawrence (1972) for a full treatment of these techniques. Many cementing materials are available and experimentation may be necessary to discover the best for a particular sediment. Low viscosity is necessary for greatest penetration, slow polymerization to prevent cracking and strains (which may give the cement a birefringence), and a strong final product is essential.

Feldspar Staining Methods

Staining is a rapid and practical means of distinguishing feldspars from each other and from other minerals in hand specimens, thin sections, and grain mounts. Details of the techniques may vary, depending on the character of the rock; experimentation is commonly necessary to obtain good results. Pores, fractures, some clays, and rarely other minerals may retain the stains—be aware and be careful! Stains are usually brittle and flaky when dry, so take precautions to avoid breaking them off.

1. Polish the rock slab (using 400 to 600-grade carborundum powder) or uncover a thin section or mount loose grains on a slide. Make sure the upper portions of the grains are exposed. With loose grains, remove any coatings first—either in an ultra-

sonic device or see "Pretreatment of Samples". For best results, apply a final polish on wet 600-grit silicon carbide paper.

2. Flush the surface of the rock slab or thin section with acetone; quickly rinse it in tap water. Immerse it in detergent solution for up to 5 min; rinse it with distilled water and air-dry.

3. Etch the surface with fresh, strong hydrofluoric acid (HF) fumes (52–55% or greater concentration; evaporation causes the acid to weaken, so check the concentration or use new acid every 45 min). The sample should be suspended 1–2 cm above the acid, preferably in a closed container. Etching time should be 25–35 sec, but this is variable depending on the strength of the acid and the character of the sample. Avoid acid condensation on the surface!

> **Caution:** HF is dangerous. Always work under a hood, and wear rubber gloves, a plastic apron, and goggles. Have neutralizing solution available (eyewash of cold saturated $MgSO_4$ solution; for skin, a paste of MgO in glycerine). Seek professional medical advice after any contact.

4. Remove the sample (tweezers work well), and immerse it immediately into an oversaturated solution of sodium-cobaltinitrite (about 5 g/10 ml distilled water. Ensure saturation is maintained by periodically shaking the solution. The shelf life is as long as 6 months if it is kept in a sealed dark glass bottle, but a fresh solution is advisable). Remove the sample after 30–45 sec (variable). K-feldspar should now show a distinctive yellow stain.

5. Rinse the sample twice in tap water, and gently shake off excess water (or blot at one end with a paper towel). The procedure ends here unless you desire to stain plagioclase feldspars.

6. Dip the sample quickly into a 2–5% solution of BaCl (indefinite shelf life). Agitate once or twice, and remove before 2 sec have elapsed. (Barium will substitute for calcium in plagioclase.)

7. Dip the sample immediately into a beaker filled with tap or distilled water; agitate it briefly, then agitate in another beaker of distilled water for 10 sec.

8. While the surface is wet, place several drops (with an eyedropper or micropipette) of potassium-

rhodizonate solution on it; tilt or agitate the sample to distribute the stain evenly. Leave for a few seconds (or until plagioclase grains are pink). (This solution is prepared by dissolving of 0.2 g rhodizonic acid dipotassium salt in 30 ml distilled water; remove coagulated clumps of powder by filtering to avoid blotchy stains. The shelf life is approximately 1 hour.) The degree of stain is proportional to calcium content in plagioclase (pure albite will not stain).

9. Rinse the sample in a beaker of water and air-dry it.

The method given is largely that of Houghton (1980). For other methods, see Wilson and Sedoran (1979), Bailey and Stevens (1960), and Hayes and Klugman (1959).

HEAVY MINERALS

Heavy minerals (HMs) are arbitrarily designated as minerals with a specific gravity greater than 2.85 (see Table 19). Whereas such minerals generally constitute only roughly 1% of the sand-size fraction of sediments, they are important indicators of provenance (ultimate source lithology) and the geological history of the sediments (weathering, transportation, and deposition). Occasionally they are useful in correlation. Before trying to solve a particular problem by heavy mineral analysis, read Rittenhouse (1943) and other relevant references from the list supplied.

Interpretations

Heavy minerals are particularly indicative of provenance, and in regional studies it is possible to recognize multiple sources contributing to single sedimentary units. Hence HMs are important tools in paleogeographic reconstruction (see for example, Rittenhouse, 1943; Rice et al. 1976). Studies of relative mineral stabilities also permit paleoclimatic interpretation (see Table 17). Paleotectonic conditions are also commonly indicated, albeit indirectly, insofar as tectonism dictates the type of source rock exposed (for example, see Stow, 1938). Detailed studies permit one to make valuable inferences about processes of transportation and deposition (see for example, Bradley, 1952; Lowright et al., 1972; Stapor, 1973; Slingerland, 1977; Flores and Shideler, 1978). Alterations after deposition, or authigenesis of HMs, indicate aspects of geochemistry during diagenesis. Lateral or vertical stratigraphic trends of HMs are particularly significant and powerful interpretive tools.

Particularly important to remember when undertaking HM studies in sediments is that because of their high specific gravity, HMs will be transported with coarser grains of the common sedimentary minerals. The hydraulic equivalence also varies between HMs—for instance, zircons, (s.g., 4.68) of one size grade will be associated with larger tourmalines (s.g., 3.0–3.25). Thus, HM ratios may vary greatly in the same sediment, depending on the size grade studied. In addition, HMs commonly occur in only a restricted size range within the source rock. You must consider these factors carefully when you plan any HM study and when you make interpretations from their occurrence.

Table 19. Classification of Minerals for Heavy Liquid and Magnetic Separation

Heavy Liquid column (vertical labels, left to right): Bromoform, Tetrabromoethane, Methylene iodide, Thallous formate, Thallous formate-malonate

Heavy Liquid	Group II		Group III		Group IV	
	2.3	Glauconite				
					2.0–2.4	Zeolites
					2.3	Gypsum
					2.5	Leucite, Kaolin
					2.5–2.6	Alkali Feldspars
					2.6–2.7	Na-Ca feldspars
					2.6–2.7	Scapolite group
					2.62	Chalcedony
	2.63	Cordierite	2.6		2.6	Nepheline
			2.9	Chlorites	2.65	Quartz
			2.8		2.72	Calcite
			3.0		2.85	Dolomite
			to	Chamosite	2.86	Phlogopite
			3.5			
				Biotites		
	3.1	Iron-rich biotite / Actinolite	3.4	Amphibole / Pyroxene	2.8–3.1	Muscovite
					3.0	Tremolite
	3.2	Iron-rich	3.0		3.3	Enstatite
	to	Amphiboles and	to	Tourmaline	3.1	Apatite
	3.6	Pyroxenes	3.25		3.2	Fluorite, Andalusite
			3.4	Epidote	3.23	Sillimanite
			3.5	Olivine (wide range of s.g.) Chloritoid	3.5	Sphene
					3.5	Topaz
	3.7	Melanite	3.65		3.6	
			to	Staurolite	to	Spinel
			3.75		3.7	
	3.8	Siderite / Limonite	3.8	Pleonaste* / Limonite	3.6	Kyanite
					3.9	Anatase / Brookite
		Garnet	4.0	Garnet (wide range s.g.)	4.0	Perovskite / Corundum
	4.0	Sphalerite				
	4.3	Almandine	4.2	Ferriferous rutile	4.2	Rutile
					3.5–4.5	Leucoxene
			4.4			
	4.4	Chromite*	to	Chrome spinel	4.5	Barite
	4.6	Xonotine	5.1		4.7	Zircon
			4.8	Pyrolusite		
	4.8	Ilmenite*	4.9	Marcasite		
	5.2	Hematite	5.0		5.0	Pyrite*
			to	Monazite		
			5.3		7.0	Cassiterite
			5.3	Scheelite	7.2–7.6	Wolframite
			to	Columbite*	14.0–19.0	Platinum
	7.0	Cassiterite	7.3		19.0–21.0	Gold

*Good conductors are noted by the asterisks and may be separated by electrostatic methods.

Note: Specific gravities given may not show the full range of the mineral.

Group I	=	Highly magnetic minerals	Magnetite 5.17 / Titanoferrite 4.65 / Pyrrhotite 4.65
Group II	=	Moderately magnetic minerals	
Group III	=	Weakly magnetic minerals (sometimes separable magnetically)	
Group IV	=	Practically nonmagnetic minerals	

Remember, the same mineral may vary considerably in magnetic properties and specific gravity according to composition, source of origin, coatings, degree of alteration, inclusions, and so on.

Source: modified from Krumbein and Pettijohn, 1938, with additions from Milner, 1962

Anomalous roundness relationships of HM grains to each other and/or to the associated light minerals are important clues to geological history. Hardness and internal anisotropism vary widely in HMs. The harder and more anisotropic grains should be more angular—if they are not, different histories can be inferred for the two types of grains. Similarly, if both rounded and angular grains of the same mineral variety occur, contributions from different source rocks (or areas) are indicated.

Zircon, tourmaline, and rutile are the most stable HMs, with respect to both physical and chemical weathering processes (see Hubert, 1962); authigenic crystals or overgrowths of these minerals are common in some sediments. Apatite and garnet are quite resistant to physical processes, but they can be rapidly destroyed under some conditions of chemical weathering or diagenesis. Sphene is very resistant to abrasion, but it commonly shows a partial alteration to leucoxene in detrital sediments; it also appears to be relatively common as minute authigenic crystals (sometimes *after* leucoxene). Other HMs (such as pyriboles and metamorphic minerals) are generally less common but are very significant for interpretations. See particularly Milner (1962, vol. 2) for occurrences and properties of the various HMs in sediments.

In HM studies, look first at the character of the entire suite present. For detailed studies it seems best to concentrate on several mineral species, looking at the same size grade for each sample. Interpret the results in terms of provenance, differential chemical and physical stability (relevant to both transportation and paleoclimatic history), hydraulic factors, and diagenetic history —and by integrating the results with interpretations from light mineral studies.

Separation Procedures

Many heavy minerals are paramagnetic, that is, they become magnetized in, and proportionately to the strength of, a magnetic field. Hence they are commonly subjected to magnetic separation after concentration using specific gravity techniques. These notes describe both separation techniques and procedures for mounting the loose grains on slides for subsequent microscopic studies.

Unfortunately, inclusions, bubbles, polycrystallinity, grain coatings, the effects of weathering, and differences inherent between species of the same pure mineral commonly result in imperfect separations: no single technique will result in a perfect separation of minerals. Experimentation is necessary with any particular suite of minerals, and samples may require several passes through the procedure before good separations are achieved.

For some studies, analysis of only one or two minerals may be desired—modifications of the following techniques may allow simple and rapid separation of these.

INITIAL PREPARATION

1. Disaggregate the sample and clean grains of any coatings that may be present. (If chemical techniques are used, be aware of the destruction or modification of minerals that may occur). See "Pretreatment of Samples."

2. Sieve the sands into the grades chosen for analysis. Choose one size grade from the class just finer than the mode of the total sand frequency distribution (heavy minerals should be most abundant in this grade) and another grade at the same size for all samples (to facilitate the comparison of suites).

3. Dry the sample.

4. Depending on the number of heavy minerals and the quantity of the sand grade for analysis, it may be best to microsplit a representative 1- to 3-g subsample.

5. Weigh the (sub) sample to 0.01 g.

HEAVY LIQUID SEPARATION

Heavy liquids of differing specific gravity are available. The most commonly used liquid is bromoform (CHBr$_3$—specific gravity of 2.85 at 20°C, changing 0.023/°C), in which the commonest light minerals (quartz, feldspar, calcite) float. Bromoform decomposes in light; therefore store it in a dark place. Alcohol, acetone, and carbon tetrachloride are soluble in bromoform and can be used in rinsing bromoform into a beaker for later recovery of the expensive heavy liquid. For other heavy liquids and solvents, see Carver (1971). Most heavy liquids can be diluted to any desired lesser specific gravity— a quick method is to add a solvent to the fluid until a pure mineral grain of the desired specific gravity "floats" in the midst of the solution.

6. Check the specific gravity of bromoform before each time you use it (use a specific gravity bottle, a special hydrometer, or the refractive index method).

7. Place the bromoform in a funnel with a pinch-clamped rubber tube or glass tap. Perform the experiment under a hood. Be sure to avoid prolonged exposure to fumes.

8. Add the sample; stir occasionally to avoid mass-trapping effects (check to make sure grains are not removed on the stirring rod). Keep a watch glass over the funnel between stirrings. Allow 10-20 min for separation.

9. Drain heavy minerals (with accompanying bromoform) into a lower funnel, lined with filter paper, leading to a storage bottle.

10. Drain the rest of the bromoform and light minerals into another funnel lined with filter paper; use the same storage bottle.

11. Replace the storage bottle with another, labeled "bromoform washings."

12. Using methyl alcohol or acetone, wash the remaining grains and bromoform from the upper funnel onto the lower filter paper, then wash the bromoform from the filter paper. Collect the alcohol and bromoform in the "washings" bottle.

13. Wash the bromoform from the filter paper containing the heavy minerals into the same "washings" bottle. The grains and filter paper should be clean of bromoform.

14. Place the filter papers and grains in a dish or beaker; dry at temperatures no greater than 50°C for 1-2 hr.

15. Weigh both lights and heavies to 0.001 g.

16. Calculate the weight percentage of the "heavies" for the size grade used.

17. Reclaim the bromoform from the "washings" bottle (see below or place a beaker containing the washings in cold brine—bromoform will freeze at 9°C and separate cleanly from admixed fluids).

Alternative to steps 7-12

This alternative is useful where few separations are necessary, grains are very small, or there are many grains with a specific gravity close to that of the heavy liquid.

7. Place 10 ml bromoform in a centrifuge tube that has a constricted end; add grains (a much smaller quantity is necessary than with the funnel method). Wash the grains into the tube with another 10 ml bromoform.

8. Balance the centrifuge (put equivalent weights in opposing tubeholders). Cork the tubes.

9. *Gradually* build up the spin of the centrifuge (there is no need to exceed approximately half-speed); hold for 5 to 15 min, then gradually reduce the spin to zero.

10. Extract the centrifuge tube(s) and follow step (a), step (b), or step (c):
(a) Insert a special stopper to seal off the lower portion of the tube with its heavies. Pour off the bromoform and light minerals through a funnel with filter paper; rinse out the remaining grains (with bromoform or with alcohol). Remove and rinse the stopper. Remove the heavies and bromoform. (A rubber policeman may be used to extract recalcitrant grains.)
(b) Insert the constricted base of the centrifuge tube into cold brine. Bromoform freezes at 9°C, and the heavies will be trapped in the frozen fluid. Extract the light minerals and fluid bromoform as in step (a), then thaw and extract the heavies and bromoform. (If frozen bromoform begins to thaw too early, refreeze the lower portion of the tube; also keep your fingers on the upper portion of the tube during the extraction of light minerals.)
(c) Use a micropipette to withdraw heavy minerals (use a pipette bulb).

Reclaiming heavy liquids

Equipment

½-gal. or 1-gal. reagent bottle
1 stopper with 2 holes for glass tubing
short length of rubber tubing
pinchcock
1 long and 1 short length of glass tubing
ring stand and ring holder
beaker
heavy liquid washings
distilled water

Procedure

1. Fit the bottle mouth with the stopper and tubing; seal carefully. The long tubing should almost reach the bottom of the bottle, and pro-

trude only 1–2 in. outside the stopper. The short tube should be flush with the inside of the stopper, and protrude outside far enough to adequately receive the rubber tubing (a glass tube with a stopcock would be even better). Place the pinchcock on the rubber tubing.

2. Place the heavy liquid washings in the bottle (no more than ¼ of the bottle's volume), and fill the bottle with distilled water (to about ¾ to ⅞ of the bottle's volume).

3. Be sure the pinchcock is on the rubber tubing. Holding an index finger over the stopper end of the long glass tube (to prevent filling it with liquid), insert the stopper firmly into the bottle top. Still holding a finger over the tubing, invert the bottle. If the bottle is properly filled, the level of the liquid in the inverted bottle will be below the open end of the long tube (thus allowing air to enter when the liquid is drained).

4. Holding the stopper securely (and still with a finger over the open end of the long tube), shake the bottle vigorously. Place the shaken bottle (still inverted on the ring stand, and allow the heavy liquid to separate from the water (1–2 hr).

5. Drain out the separated heavy liquid through the rubber tubing (taking care not to drain out any water), rinse the bottle thoroughly with water, and repeat the process four or five times.

6. When reclamation is complete (heavy liquid should be clear in contrast to its earlier cloudy appearance), filter the heavy liquid through several thicknesses of filter paper. Reclamation will not normally bring the specific gravity of the heavy liquid back to its original value, but it should be within 0.05 of the original.

ELECTROMAGNETIC SEPARATION

Electromagnetic separation is applicable either to the original dry sample (the method applies to both light and heavy minerals) or to the heavy mineral separate (separation into subclasses will make mineral identification easier).

6. Remove ferromagnetic grains with a hand magnet (with thin paper between the magnet and grains). Weigh and store separately. Do **not** allow these grains into the magnetic separator.

7. Feed the grains into a Frantz isodynamic magnetic separator. Set the forward and side slopes as required (see Fig. 83; it may be best to begin with 15° side, 25° forward tilt). Set the milliamperage at a low value (0.3 mA first), then increase as required for the following separations. (For the cleanest separation, the grains have to be run through at one setting several times.)

Do not have a watch or other magnetizable device in the room while the Frantz is working.

Mounting Loose Grains

1. Sieve and clean the grains (see separate sections).

2. If a 0.03-mm section is desired, warm Lakeside cement to 88–99°C and apply a thin coating to a microscope slide over an area about 1½ times the width of the cover glass.

3. Cool the slide, and then reheat to 101°C (this helps eliminate bubbles).

4. Scatter grains evenly across the cement; take care that a representative suite of grains is sampled (some segregation by shape or density may occur within the container).

5. Cool the slide and grind with 400 and then 600-grade carborundum paste.

6. Reheat, apply a thin layer of cement, and carefully press down a coverslip (using, for instance, the end of a pencil eraser) until bubbles are eliminated and excess cement is forced out around the edges (creating an air seal).

7. Cool and label the slide.

If an uncovered mount is desired for etching and staining of minerals, follow the above procedure steps 1 to 4 or 5, and stain before step 6 (if a coverslip is necessary). Or,

2. Apply a very thin coating of canada balsam to the slide (one drop spread over half the area of the slide).

3. Heat 3–4 min until the balsam is cooked.

4. Cool 1–3 min, and then spread the grains evenly on the slide.

5. Place the slide on a hotplate at 65°C and remove it when the grains begin to sink (no grain should be covered by balsam).

6. Cool and label. (Excess balsam can be removed with alcohol or acetone.)

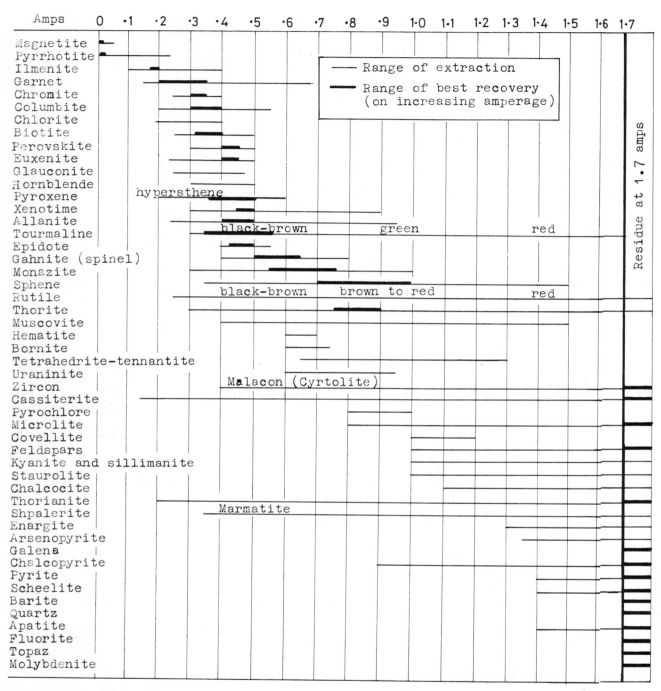

Figure 83. Magnetic susceptibilities of minerals in Frantz Isodynamic Magnetic Separator (side tilt 15°, forward tilt 25°, grains 100–150 μ). *(After Rosenblum, 1958)*

Or:

2. Prepare a plastic (epoxy) compound, and allow it to stand until bubbles disappear.

3. Spread liquid over part of the glass slide (if you spread it too thinly, the plastic will blister when heated).

4. Dry for 12 hr at 70–80°C.

5. Mix more plastic (for 3∅ grains, use one part plastic to three parts thinner; use less thinner with larger grains).

6. Using a fine brush, paint the liquid across the hardened plastic, then sprinkle grains on the slide, invert the slide to remove loose grains, and dry it for 12 hr at 90°C.

7. Cool and label.

If a temporary mount is desired for immersion oil methods of mineral identification (after the method of Marshall and Jeffries, 1946), follow these four steps:

1. Prepare solution A: 0.1 gelatin solution (¼g/50 ml H₂O). Prepare solution B: 10 ml distilled water, 5 ml acetone, 2 ml 2% formalin. Cork both well to avoid evaporation.

2. Place drops of solution A on a glass slide; dry at 80°C. (For large grains, use more gelatin).

3. Place one or two drops of solution B on the gelatin film, spread the grains evenly on the soft surface with a needle. Dry at 80°C.

4. Place refractive index oils on the slide as necessary; between oil applications remove the oils with acetone (xylol, alcohol, or carbon tetra-chloride) and dry.

CLAY MINERALS

Clay minerals are hydrated aluminophyllosilicates smaller than 0.004 mm (8∅) in size (according to the Wentworth scale); rarely, larger crystals occur (for example, macro-kaolinite). In fact, most clay minerals are finer than 2μ, which is the upper limit for the clay fraction in pedologists' usage and is a widely accepted standard. The most abundant sediment, they are formed either by weathering or diagenetic processes. During diagenesis, clay minerals commonly recrystallize and grow to form larger authigenic micas or chlorites, and it can be difficult to distinguish them from detrital flakes. All the clays are biaxial negative with a small 2V and length slow; hence they are difficult to differentiate in thin section, particularly because aggregation effects (such as preferred fabrics or stacking of flakes in the rock slice) or stains (such as those of iron oxides) may change their appearance. X-ray diffraction (XRD), infrared spectroscopy (IR), or differential thermal analysis (DTA) must generally be applied to determine the mineralogy of the clays. Table 20 and Figs. 84–88 illustrate the crystallography, composition, classification, selected XRD/DTA patterns, and origin of clay minerals (also see Grim, 1968; Millot, 1970; Brindley and Brown, 1980). Note that modern usage is tending to recognize a Smectite Group instead of a Montmorillonite Group; the Smectite Group is subdivided into montmorillonite (Mg substituting for Al in the octahedral sheet) and beidellite (Al substituting for Si in the tetrahedral sheet) subgroups.

Even though identification may require XRD or DTA, petrographic analysis is still essential to determine textural relationships, which may be critical to the interpretation of the origin of the clay minerals (for example, detrital versus early diagenetic versus late diagenetic). Under Crossed Nicols, sericite (fine muscovite) has the highest birefringence and first reds and blues are common; illite (hydro-muscovite) grades into sericite, but most commonly shows first-order whites and yellows; kaolin has a very low gray birefringence, and is difficult to distinguish from chlorite (which has equally low birefringence—anomalous birefringence if you're lucky—and takes on a greenish color in plane-polarized light if it's thick enough). Montmorillonite also has a very low birefringence, but it is the only common clay mineral with a refractive index below that of balsam; the main difficulty may be to distinguish it from authigenic interstitial chert.

Table 20. Basis of Classification of Major Clay Mineral Groups

Combination of structural units—unit layers of clays formed of:

    1(silica tetrahedral sheet):1(alumina octahedral sheet)—Kaolinite Group
    1:2—Montmorillonite and Illite Groups
    1:2 + 1(brucite sheet)—Normal (sedimentary) Chlorite Group
    (1:1 + 1—Septechlorite Group, of metamorphic origin)

Cation content of octahedral sheet:

    Dioctahedral—$2/3$ of all possible cation lattice positions filled, as for gibbsite
        $Al_2(OH)_6$ (glauconite, most illite and montmorillonite, muscovite)
    Trioctahedral—all possible cation lattice positions filled, as for brucite
        $Mg_3(OH)_6$ (vermiculite, normal chlorites, phlogopite, biotite)

Manner and perfection of stacking layers:

    The vertical stacking array of clay layers, such as kaolinite-illite-kaolinite-
        illite-etc. vs. illite-montmorillonite-chlorite-montmorillonite-illite-etc.
    The crystallographic symmetry of the vertical array of clay layers, such as
        2M (two-layer monoclinic configuration), 1M (one-layer monoclinic),
        1 md (one-layer monoclinic disordered), and 3T (three-layer trigonal)

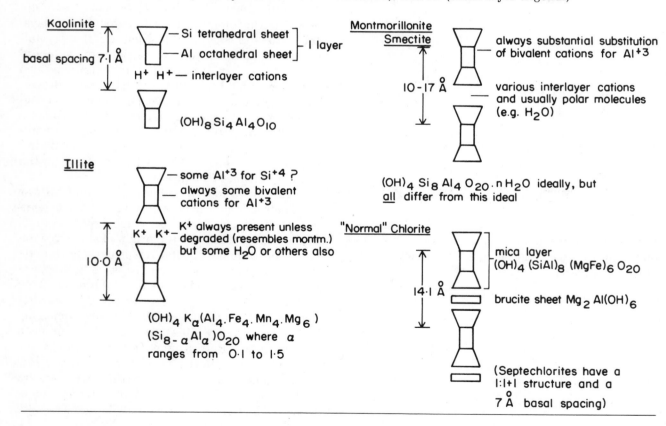

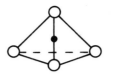

Silica tetrahedron
Si^{4+} central and equidistant from 4 O^{2-} ions; units link laterally to form a sheet 4.93Å thick (Al^{3+} substitutes for some Si^{4+} in most clays)

linked
vertically
with
sheets
of

Alumina octahedra
Al^{3+} central and equidistant from 6 O^{2-} ions, units link laterally to form a sheet 5.05Å thick (Mg^{2+} substitutes for some Al^{3+}, and other bivalent ions for Mg^{2+}; $(OH)^-$ substitutes for some O^{2-})

Brucite sheets: octahedral units of $MG_2(OH)_6$, laterally linked to form separate sheets in chlorites.

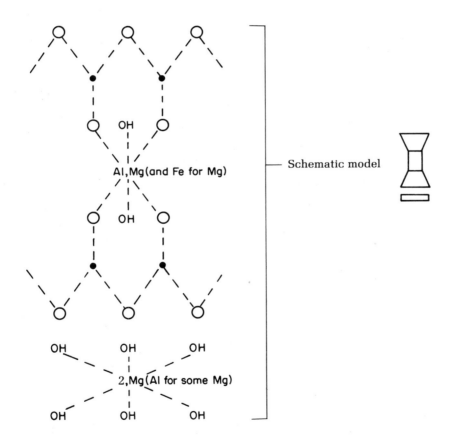

Figure 84. Fundamental structure of clay minerals. *Exceptions:* amorphous clay minerals (allophanes)—associations of tetrahedral and octahedral units in random arrangements; and double-chain minerals (palygorskite group)—chains of silica tetrahedra in amphibole-like structures linked with octahedral groups.

153

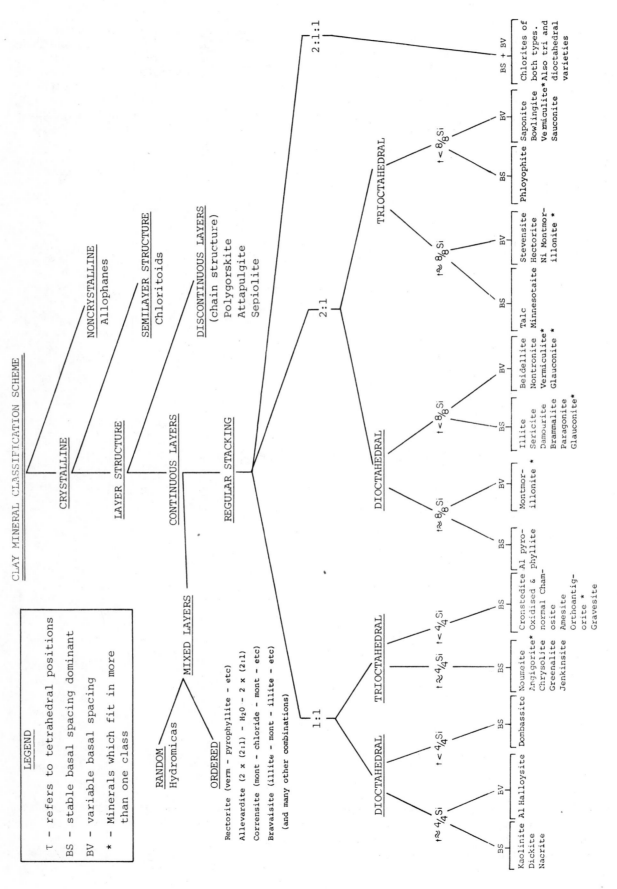

Figure 85. A clay mineral classification scheme by D. M. McConchie (1978).

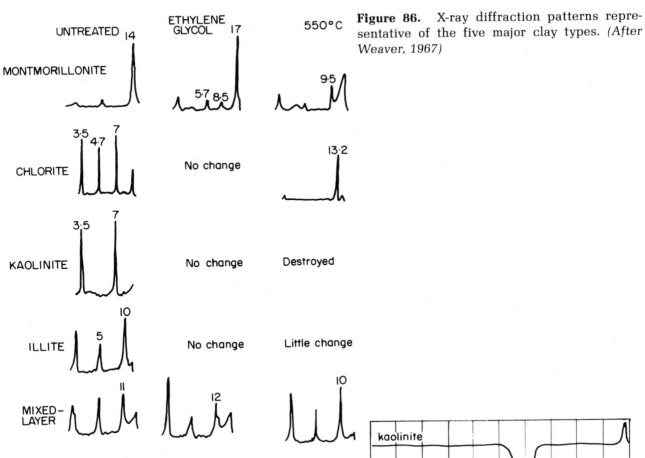

UNTREATED

ETHYLENE GLYCOL

550°C

MONTMORILLONITE

14

17

5·7 8·5

9·5

CHLORITE

3·5 4·7 7

No change

13·2

KAOLINITE

3·5 7

No change

Destroyed

ILLITE

5 10

No change

Little change

MIXED-LAYER

11

12

10

Figure 86. X-ray diffraction patterns representative of the five major clay types. *(After Weaver, 1967)*

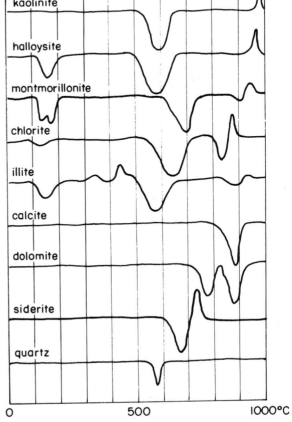

kaolinite

halloysite

montmorillonite

chlorite

illite

calcite

dolomite

siderite

quartz

0   500   1000°C

Figure 87. DTA curves of some important sediment-forming minerals (heating rate, 10°C/min). Peaks in angstroms (Å). *(After Muller, 1967)*

GENETIC MODEL

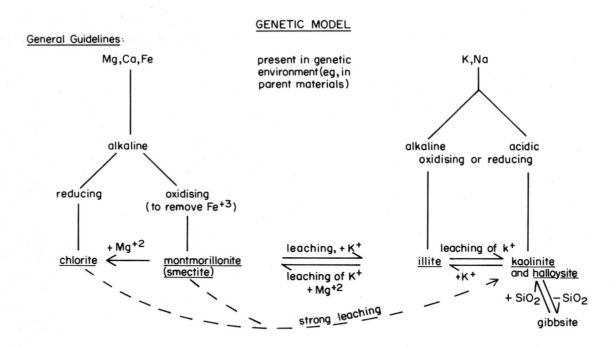

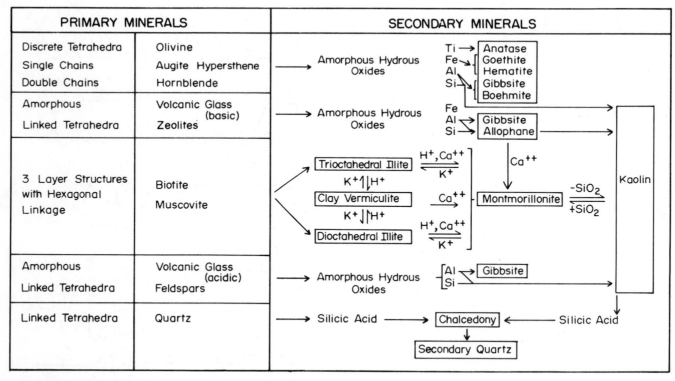

Figure 88. Genesis of clay minerals. *Primary clays of neoformation*—derived from nonclay solids, such as other minerals and volcanic glass, or from ions and colloids in aqueous solution and suspension. *Transformation clays* (clays of N + 1 stage)—derived from other clay minerals through a fundamental crystallographic rearrangement (not merely cation exchange phenomena). *Weathering products of primary silicates* are shown in the tabular portion of the figure *(after Fieldes and Swindale, 1954).*

Allophane, an amorphous clay, is isotropic and generally requires IR analysis for identification. The double-chain clay minerals (Palygorskite Group) are very difficult to distinguish in thin section (for example, from zeolites) but are generally rare. Many mixed-layer clays are common and cannot be distinguished without XRD, IR, or DTA.

Clay minerals may be precipitated directly from aqueous solutions or suspensions during diagenesis (see Wilson and Pittman, 1977), or they may form as alteration products of silicate minerals during initial weathering or diagenesis. Formation (or alteration) by weathering after the sediments being studied were brought into the modern weathering environment (by uplift and erosion) is also possible, particularly in permeable rocks. Which clay mineral forms depends on the chemistry of the weathering or diagenetic environment, influenced by source rock composition and both macro- and microclimatic factors (see Singer, 1980). The influence of biota and organic decomposition products is commonly of great importance (for example, bacterial activities; degradation of minerals as they pass through the alimentary canals of sediment-feeders; organic chelating compounds, acids, and bases). Also critical are the type and concentration of inorganic ions in the genetic environment (derived from the decomposition of minerals and supplied by migrating interstitial waters). The time factor is important —products may be similar (irrespective of the parent rock composition) when chemical activity is intense over a short time span or when it is sluggish but prolonged. Clay minerals may be produced or modified in the marine environment (the term *halmyrolysis* refers to changes taking place on the sea floor) as well as in nonmarine settings. Only clays of the Kaolin Group cannot be formed in the marine setting, because they require acidic leaching conditions (beidellite—an aluminum-rich smectite—is also unlikely to form in magnesium-rich seawater). Chlorite and montmorillonite form under alkaline conditions where leaching is restricted and where there are supplies of magnesium (and, to a lesser degree, iron and calcium) ions. Most sedimentary chlorites appear to be products of diagenesis, often from montmorillonite but also from ferromagnesium minerals. Illite forms where there are supplies of potassium ions, and since such ions may be supplied from the breakdown of minerals such as K-feldspars and micas, it is characteristic of weathering/diagenetic products derived from acidic igneous/metamorphic rocks, whereas the chlorites/montmorillonites are characteristic of products from more basic rocks. Glau-

conite, a clay mineral of the Illite Group, is always a halmyrolitic or diagenetic product, formed where potassium ions (for example, from seawater) and iron ions are available.

Clay minerals may also be "transformed" from one variety to another during diagenesis (see, for example, de Segonzac, 1970). Arguments rage as to how many fundamental crystallographic transformations take place in this way; for example, Weaver (1967) presents arguments against substantial widespread changes, whereas many other workers argue for such changes. Vertical changes in clay mineralogy within some stratigraphic piles, wherein other effects can be discounted, strongly indicate that late-stage diagenesis can cause crystallographic transformations, particularly of montmorillonites through mixed-layer clays to illites and chlorites. Such transformations appear to be particularly important as a late-stage means of producing water to flush out petroleum hydrocarbons. However, early diagenetic or halmyrolitic crystallographic transformations may be rare—progressive changes in clay mineralogy laterally, such as in an offshore direction as noted within modern sediments, often reflect merely selective sorting of the different clays supplied (which have different crystallite sizes and/or behave differently in ion-rich seawater). Degradation and regradation, affecting interlayer cations, are common transformations at any stage in the sediment history.

Complicating investigations still further, clay minerals may of course be recycled from older sedimentary rocks. Whether those in the new deposit are recycled or derived from weathering of other primary minerals in the source area is commonly indeterminate. Much work and experience are generally necessary to determine the origin of clays in sediments.

Glauconite

Glauconite constitutes a specific clay mineral that is particularly useful in geological interpretations. Most commonly, it occurs as rounded green pellets of sand size (100–1000 μm) in sandstones (arenites, limestones, greensands), but it also occurs (rarely) as a coating on grains or on broad substrates (such as hardgrounds). Distinctive morphological varieties can be determined from binocular, petrographic, or scanning electron microscopy (see Triplehorn, 1966, and McConchie and Lewis, 1980; also Table 21). X-ray diffraction (XRD) is necessary to determine mineralogical varieties and to distinguish

Table 21. Varieties of Glauconite

Morphological varieties (determined with binocular microscope):

Ovoid and spheroidal pellets

Lobate pellets (deeply mammillated surfaces; lobes tend to break off during transport, hence presence of this morphology is best indication of authigenic glauconite)

Tabular and discoidal pellets

Vermicular pellets (due to replacement of micas or some types of fecal pellets)

Fossil casts and internal molds

Composite grains

Fragmentary grains: subclass A if more angular than 0.5 on roundness scale, probably reflecting breakage during perigenic transport; subclass B if more rounded than 0.5, probably reflecting abrasion after breakage and detrital (recycled or allogenic) origin

Spongy pellets: porous, earthy surface form. Using scanning electron microscopy (SEM), *cauliflower* vs. *serrulate* varieties can be distinguished according to whether protuberances and pores are rounded and indistinct or sharply serrated and distinct. Spongy pellets appear to reflect diagenetic corrosion of other types of glauconite pellets.

Varieties of internal texture (determined in thin section):

Random microcrystalline (crystallites in all orientations)

Oriented microcrystalline (crystallite orientation produces straight or wavy extinction of grain UXN)

Patch-oriented microcrystalline (patches of grains have oriented crystallites, remainder is random microcrystalline)

Micaceous

Coatings

Fibroradiated rims (coating has microflakes oriented perpendicular to rim)

Organic replacements

Crystallographic varieties (determined by XRD): Defined using % expandable layers present and disorder coefficient, $DC = \tau b/(h-b)$, where τ (measured for convenience in degrees 2θ) = half-height line width of 10 Å XRD peak for an oriented sample previously heated to 400°C for 1 hr; h = intensity of the 10 Å peak; b = background intensity (both h and b have same arbitrary-length units).

Well-ordered glauconite: $DC \leq 0.25$.
   Sharp, symmetrical XRD peaks will be apparent. Will comprise most glauconites with <10% expandable layers.

Disordered glauconite
   1. Moderately disordered—$0.25 < DC \leq 0.5$ and $\leq 40\%$ expandable layers. XRD pattern may show peak asymmetry.
   2. Extremely disordered—$DC > 0.5$ and $\leq 40\%$ expandable layers. XRD pattern will generally show pronounced peak asymmetry.

Interlayered glauconite: > 40% expandable layers; only 1 distinct mineral type apparent.

Mixed-mineral glauconite: 2 or more distinct mineral varieties present, only 1 of which is glauconite (one of the previous classes).

Source: after Triplehorn, 1966; McConchie and Lewis, 1978, 1980

158

glauconite from other clay minerals (see Fig. 89). McRae (1972) and Odin and Matter (1981) provide excellent reviews of existing knowledge (also see McConchie and Lewis, 1978, 1980; McConchie et al., 1979).

Glauconite minerals are iron- and potassium-rich hydrated aluminophyllosilicates with a 2:1 layer lattice. There is a wide range in expandable layer content within the mineral family; according to Odin and Matter (1981), there is a fully gradational suite from *glauconitic smectite* (disordered, high expandable layer content) to *glauconitic mica* (well ordered, less than 5% expandables), reflecting progressive evolution in the process of glauconitization. Others subdivide the family into a series of crystallographic varieties shown in Table 21 (see Burst, 1958; McConchie and Lewis, 1980). The term *mineral glauconite*, or *glauconite* alone, has been used in the sense of the well-ordered end-member, but *glauconite* has also been used in a broad sense to avoid confusion, Odin and Matter proposed that *glaucony* be used in the general sense for all glauconitic minerals. Total iron content is greater that 15% (most have 19-27%) with a $Fe^{2+}:Fe^{3+}$ ratio of approximately 1:7; K_2O content is greater than 3% (up to 9%). Al_2O_3 generally ranges from 5-8%, and MgO from 3-5%. Electron exchange occurs between adjacent Fe^{2+} and Fe^{3+} ions at temperatures above about 80 K (McConchie et al., 1979). Some grains contain both glauconitic and nonglauconitic minerals; these are distinguished morphologically as *composite grains* and/or mineralogically as *mixed mineral glauconite*. Table 21 describes the different varieties of glauconite.

Greenalite (essentially Precambrian) and *chamosite*, also called *berthierine* (Precambrian and Phanerozoic), resemble glauconite under the microscope (some chamosite is distinctive as oölites). *Celadonite* (an alteration product or vug filling of some basic volcanics) is indistinguishable from glauconite except by its occurrence or by complex instrumental methods such as Mossbauer analysis. Chamosite is most commonly misidentified as glauconite; however, its XRD pattern is different (e.g., a nearly constant 7.5 Å basal spacing).

ORIGIN OF GLAUCONITE

Glauconite pellets are formed in the marine environment at the sediment-water interface, as are glauconitic films or coatings; some interstitial glauconitic cements originate below the interface. Non-marine glauconite has been reported, but it differs chemically from the glauconitic mineral family defined here (see Kossovskaya and Drits, 1970; Odin and Matter, 1981). The marine glauconites are forming today and occur in rocks as old as 2×10^9 years. They are most abundant today in the outer-shelf and upper-slope environments (water depths of 50-500 m), but it is unlikely that water depth has any control, and pellets can be redeposited and may form in shallower or deeper environments. A slow net sediment accumulation rate is essential —Odin and Matter (1981) estimate necessary exposure times at the sediment-water interface to be 10^3-10^4 years for initial glauconitization, and 10^5-10^6 years for development of highly evolved glauconitic mica. Slightly reducing conditions (for supplies of Fe^{2+}) but an absence of H_2S (sulfide ions preferentially scavenge iron) are necessary. Once formed, however, the pellets can be and are commonly redeposited rapidly in energetic, well-oxygenated settings (for example, Ward and Lewis, 1975). Most pellets are probably perigenic (locally redeposited); some are allogenic (recycled from older sediments, in which case they are few and generally fragmental— see McConchie and Lewis, 1978), and others are authigenic (*in situ*).

Glauconite minerals form as internal molds of microfossil tests and replace a variety of mineral grains and rock fragments—most commonly fecal pellets, but also quartz, pyriboles, volcanic fragments, bioclasts, and others. Odin and Matter (1981) recognize four evolutionary stages: (1) precipitation as (micro-)pore fillings, initial substrate present and dominant; (2) initial substrate disappearing, with grains becoming green and developing paramagnetic properties of glaucony; (3) initial substrate essentially gone, with differential recrystallization and crystalline growth expanding original grain or void to produce bulbous grains with surficial cracks; (4) evolution to glauconitic mica, with new glauconite mineral forming in cracks to produce a smooth crust and rounded, subspherical shape. Others argue that glauconite forms by adsorption of iron, then potassium, into an initially disordered, swelling clay mineral lattice that may be inherited from a pre-existing mineral or be an authigenic precipitate itself. Further work is necessary; both explanations may apply.

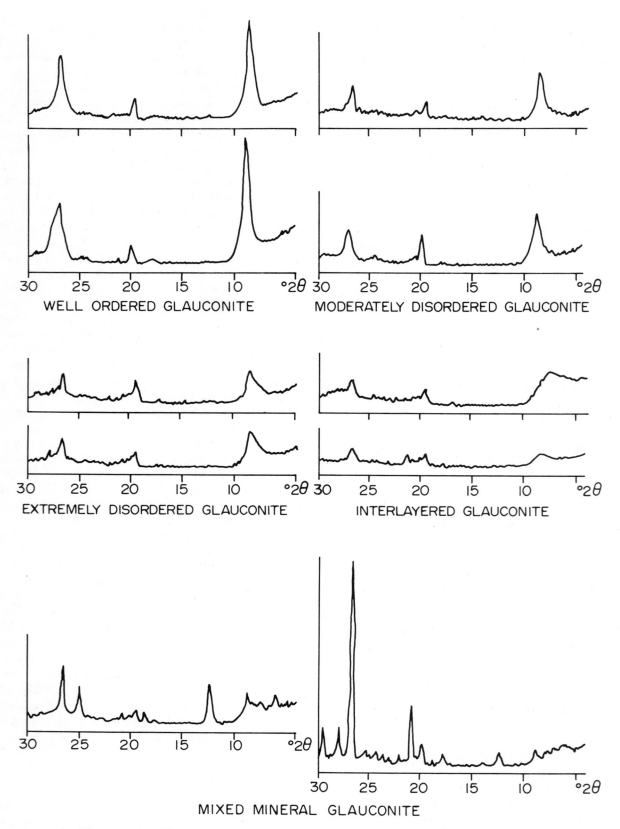

Figure 89. Glauconite analysis. Examples of XRD patterns for the crystallographic varieties of glauconite. *(After McConchie and Lewis, 1980)*

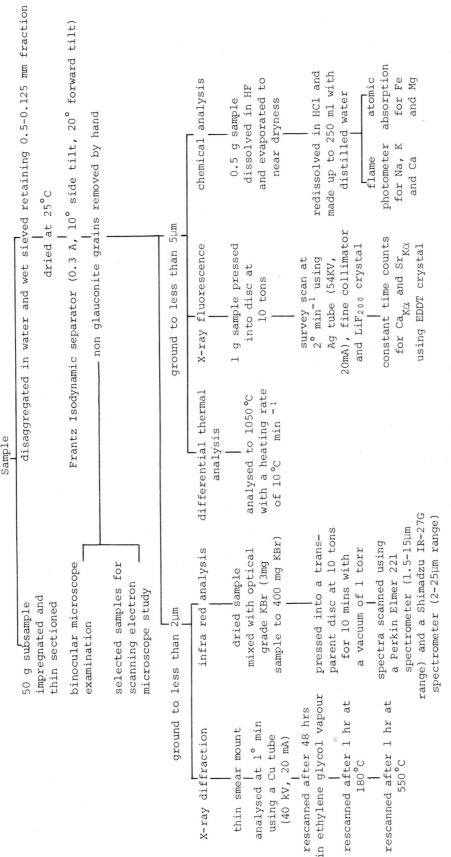

Figure 89. (Continued) Example of flow sheet for analysis of glauconites. (After McConchie and Lewis, 1978)

LITERATURE

Arenites

Alimen, H., 1965, The Quaternary Era in the Northwest Sahara, *Geol. Soc. America Spec. Paper* **84**:273-291.

Allman, M., and D. F. Lawrence, 1972, *Geological Laboratory Techniques*, Arco Publishing, New York, 335p.

Bailey, E. H., and R. E. Stevens, 1960, Selective staining of K-feldspar and plagioclase on rock slabs and in thin sections, *Am. Mineralogist* **45**:1020-1027.

Basu, A., 1976, Petrology of Holocene fluvial sand derived from plutonic source rocks: implications to paleoclimatic interpretation, *Jour. Sed. Petrology* **46**:694-709.

Basu, A., S. W. Young, L. J. Suttner, W. C. James, and G. H. Mack, 1975, Re-evaluation of the use of undulatory extinction and polycrystallinity in detrital quartz for provenance interpretation, *Jour. Sed. Petrology* **45**:873-882.

Blatt, H., 1967a, Provenance determinations and recycling of sediments, *Jour. Sed. Petrology* **37**:1031-1044.

Blatt, H., 1967b, Original characteristics of clastic quartz grains, *Jour. Sed. Petrology* **37**:401-424.

Blatt, H., and J. M. Christie, 1963, Undulatory extinction in quartz of igneous and metamorphic rocks and its significance in provenance studies of sedimentary rocks, *Jour. Sed. Petrology* **33**:559-579.

Boggs, S. Jr., 1968, Experimental study of rock fragments, *Jour. Sed. Petrology* **38**:1326-1339.

Boswell, P. G. H., 1960, The term Greywacke, *Jour. Sed. Petrology* **30**:154-157.

Brewer, R., 1964, *Fabric and Mineral Analysis of Soils*, Wiley, New York, 470p. (Reprinted in 1976 by Robert E. Kreiger Publishing, Huntington, N.Y., 482p.)

Breyer, J. A., and H. A. Bart, 1978, The composition of fluvial sands in a temperate semi-arid region, *Jour. Sed. Petrology* **48**:1311-1320.

Carozzi, A. V., 1960, *Microscopic Sedimentary Petrography*, Wiley, New York, 405p.

Conolly, J. R., 1965, The occurrence of polycrystallinity and undulatory extinction in quartz of sandstones, *Jour. Sed. Petrology* **35**:116-135.

Crook, K. A. W., 1967, Tectonics, climate, and sedimentation, 7th Sedimentological Congress.

Cummins, W. A., 1962, The greywacke problem: *Liverpool Manchester Geol. Jour.* **3**:51-72.

Davies, D. K., and F. G. Ethridge, 1975, Sandstone composition and depositional environment: *Am. Assoc. Petroleum Geologists Bull.* **59**:239-264.

Dickinson, W. R. (ed.), 1974, *Tectonics and Sedimentation*, SEPM Spec. Pub. No. 22, Society of Economic Paleontologists and Mineralogists, Tulsa, Okla., 204p. (See esp. paper by Dickinson, pp. 1-27.)

Dickinson, W. R., and C. A. Suczek, 1979, Plate tectonics and sandstone compositions, *Am. Assoc. Petroleum Geologists Bull* **63**:2164-2182.

Dott, R. H., Jr., 1964, Wacke, greywacke and matrix— what approach to immature sandstone classification? *Jour. Sed. Petrology* **34**:625-632.

Dott, R. H. Jr, 1978, Tectonics and sedimentation a century later, *Earth-Sci. Rev.* **14**:1-34.

Folk, R. L., 1974, *Petrology of Sedimentary Rocks*, Hemphill, Austin, Tex., 182p.

Folk, R. L., P. B. Andrews, and D. W. Lewis, 1970, Detrital sedimentary rock classification and nomenclature for use in New Zealand, *New Zealand Jour. Geology and Geophysics* **13**:937-968.

Gilligan, A., 1919, The petrology of the Millstone Grit of Yorkshire, *Geol. Soc. London Jour.* **75**:251-292.

Greensmith, J. T., 1963, Clastic quartz, provenance and sedimentation, *Nature* **197**:345-347.

Halley, R. B., 1978, Estimating pore and cement volumes in thin section, *Jour. Sed. Petrology* **48**:642-650.

Hayes, J. R., and M. A. Klugman, 1959, Feldspar staining methods, *Jour. Sed. Petrology* **29**:227-232.

Houghton, H. F., 1980, Refined techniques for staining plagioclase and alkali feldspars in thin section, *Jour. Sed. Petrology* **50**:629-631.

James, N. C., G. H. Mack, and L. J. Suttner, 1981, Relative alteration of microcline and sodic plagioclase in semi-arid and humid climates, *Jour. Sed. Petrology* **51**: 151-163.

Keller, W. D., and R. F. Littlefield, 1950, Inclusions in the quartz of igneous and metamorphic rocks, *Jour. Sed. Petrology* **20**:74-84.

Klein, G. de V., 1963, Analysis and review of sandstone classifications in the North American geological literature, 1940-1960, *Geol. Soc. America Bull.* **74**:555-575.

Krynine, P. D., 1940, Petrology and genesis of the Third Bradford Sand, *Pennsylvania State Coll. Min. Ind. Exp. Sta. 29*, 134p.

Krynine, P. D., 1942, Differential sedimentation and its products during one complete geosynclinal cycle, *1st Pan. Am. Cong. Min. Engr. Geol. Annals*, Mexico, vol. 2, part 1, 537-561.

Krynine, P. D., 1948, The megascopic study and field classification of sedimentary rocks, *Jour. Geology* **56**:130-165.

Lewis, D. W., 1964, "Perigenic"; a new term, *Jour. Sed. Petrology* **34**:875-876.

Lewis, D. W., 1971, Qualitative petrographic interpretation of Potsdam Sandstone (Cambrian) Southwest Quebec, *Can. Jour. Earth Sci.* **8**:853-882.

Lyell, C., 1837, *Principles of Geology*, vol. 1, 5th ed. John Murray, London, 462p.

McBride, E. F., 1963, A classification of common sandstones, *Jour. Sed. Petrology* **33**:664-669.

Mack, G. H., 1978, The survivability of labile light mineral grains in fluvial, aeolian and littoral marine environments: the Permian Cutler and Cedar Mesa Formations, Moab, Utah, *Sedimentology* **25**:587-604.

Mack, G. H., 1981, Composition of modern stream sand in a humid climate derived from a low-grade meta-

morphic and sedimentary foreland fold-thrust belt of
north Georgia, *Jour. Sed. Petrology* **51**:1247-1258.

Mitchell, A. H. G., and H. G. Reading, 1978, Sedimentation and tectonics, in H. G. Reading (ed.), *Sedimentary
Environments and Facies*, Blackwell, Boston, 439-476.

Moss, A. J., 1966, Origin, shaping and significance of
quartz sand grains, *Geol. Soc. Australia Jour.* **13**:97-136.

Oriel, S. S., 1949, Definitions of arkose, *Am. Jour. Sci.*
247:824-829.

Pettijohn, F. J., 1943, Archean sedimentation, *Geol. Soc.
America Bull.* **54**:925-972.

Pettijohn, F. J., P. E. Potter, and R. Siever, 1972, *Sand
and Sandstone*, Springer-Verlag, New York, 618p.

Pittman, E. D., 1970, Plagioclase feldspars as an indicator
of provenance in sedimentary rocks, *Jour. Sed. Petrology* **40**:591-598.

Scholle, P. A., 1979, A color illustrated guide to constituents, textures, cements and porosities of sandstones
and associated rocks, *Am. Assoc. Petroleum Geologists
Mem. 28*, 201p.

Shukis, P. S., and F. G. Ethridge, 1975, A petrographic
reconnaissance of sand size sediment, upper St. Francis
River, Southeastern Missouri, *Jour. Sed. Petrology*
45:115-127.

Suttner, J. J., A. Basu, and G. H. Mack, 1981, Climate
and the origin of quartz arenites, *Jour. Sed. Petrology*
51:1235-1246.

Todd, T. W., 1968, Paleoclimatology and the relative stability of feldspar minerals under atmospheric conditions, *Jour. Sed. Petrology* **38**:832-844.

Van der Plas, L., 1966, The Identification of Detrital
Feldspars, Developments in Sedimentology 6, Elsevier,
Amsterdam, 305p.

Williams, G. E., 1968, Torridonian weathering and its
bearing on Torridonian paleoclimate and source, *Scot-
tish Jour. Geol.* **4**:164-184.

Wilson, M. D., and S. S. Sedora, 1979, An improved thin
section stain for potash feldspar, *Jour. Sed. Petrology*
49:637-638.

Wolf, K. H., 1971, Textural and compositional transitional
stages between various lithic grain types (with a com-
ment on "interpreting detrital modes of greywacke and
arkose"), *Jour. Sed. Petrology* **41**:328-332, 889.

Young, S. W., 1976, Petrographic textures of detrital polycrystalline quartz as an aid to interpreting crystalline
source rocks, *Jour. Sed. Petrology* **46**:595-603.

Rudites

Abbott, P. L., and G. L. Peterson, 1978, Effects of abrasion durability on conglomerate clast populations: examples from Cretaceous and Eocene conglomerates of
the San Diego area, California, *Jour. Sed. Petrology*
48:31-42.

Boggs, S. Jr., 1969, Relationship of size and composition
in pebble counts, *Jour. Sed. Petrology* **39**:1243-1246.

Crowell, J., 1957, Origin of pebbly mudstones, *Geol. Soc.
America Bull.* **68**:993-1009.

Harms, J. C., J. B. Southard, D. R. Spearing, and R. G.
Walker, 1975, Conglomerate-sedimentary structures
and facies models, in *Depositional Environments as
Interpreted from Primary Sedimentary Structures and
Stratification Sequences*, Soc. Econ. Paleontologists
and Mineralogists Short Course No. 2, 133-161.

Pettijohn, F. J., 1975, *Sedimentary Rocks*, 3rd ed., Harper
& Row, New York, pp. 154-194.

Twenhofel, W. H., 1947, The environmental significance
of conglomerates, *Jour. Sed. Petrology* **17**:199-128.

Lutites

Flawn, P. T., 1953, Petrographic classification of argillaceous sedimentary and low grade metamorphic rocks
in subsurface, *Am. Assoc. Petroleum Geologists Bull.*
37:560-565.

Krumbein, W. C., 1947, Shales and their environmental
significance, *Jour. Sed. Petrology* **17**:101-108.

Picard, M. P., 1971, Classification of fine-grained sedimentary rocks, *Jour. Sed. Petrology* **41**:179-195.

Potter, P. E., J. B. Maynard, and W. A. Pryor, 1980, *Sedimentology of Shale*, Springer-Verlag, New York, 306p.

Shaw, D. B., and C. E. Weaver, 1965, The mineralogical
composition of shales, *Jour. Sed. Petrology* **35**:213-222.

Tourtelot, H. A., 1960, Origin and use of the word shale,
Am. Jour. Sci. **258A**:335-343.

Yaalon, D. H., 1962, Mineral composition of the average
shale, *Clay Minerals Bull.* **5**:31-36.

Heavy Minerals

Baker, G., 1962, *Detrital Heavy Minerals in Natural
Accumulates*, Australasian Inst. Mining and Metallurgy Monograph Series No. 1, 146p.

Blatt, H., and B. Sutherland, 1969, Intrastratal solution
and non-opaque heavy minerals in shales, *Jour. Sed.
Petrology* **39**:591-600.

Bradley, J. S., 1952, Differentiation of marine and sub-
aerial sedimentary environments by volume percentage of heavy minerals, Mustang Island, Texas, *Jour.
Sed. Petrology* **27**:116-125.

Carver, R.E., 1971, *Procedures in Sedimentary Petrology*,
Wiley-Interscience, New York, chapters by Carver,
427-452, and J. F. Hubert, 453-478.

Cohee, G. V., 1937, Inexpensive equipment for reclaiming heavy liquids, *Jour. Sed. Petrology* **7**:34-35.

Dryden, L., and C. Dryden, 1946, Comparative rates of
weathering of some common heavy minerals, *Jour.
Sed. Petrology* **16**:91-96.

Fairburn, H. W., 1955, Concentration of heavy acces-
sories from large rock samples, *Am. Mineralogist*
40:458-468.

Flores, R. M., and G. L. Shideler, 1978, Factors controlling heavy-mineral variations on the South Texas outer continental shelf, Gulf of Mexico, *Jour. Sed. Petrology* **48**:269-280.

Force, E. R., 1980, The provenance of rutile, *Jour. Sed. Petrology* **50**:485-488.

Füchtbauer, H., J. P. Milliman, A. P. Lisitzyn, and E. Seibold, (eds.), 1973, *Contributions to Sedimentology*, vol. 1: *Stability of Heavy Minerals*, 125p.

Gow, A. J., 1967, Petrographic studies of ironsands and associated sediments near Hawera, South Taranaka, *New Zealand Jour. Geology and Geophysics* **10**:675-695.

Gunn, C. B., 1968, Field concentration of heavy minerals, *Jour. Sed. Petrology* **38**:1362.

Henningsen, D., 1967, Crushing of sedimentary rock samples and its effect on shape and number of heavy minerals, *Sedimentology* **8**:253-255.

Hubert, J. F., 1962, A zircon-tourmaline-rutile maturity index and the interdependence of the composition of heavy mineral assemblages with the gross composition and texture of sandstones, *Jour. Sed. Petrology* **32**:440-450.

Hunter, R. E., 1967, A rapid method for determining weight percentages of unsieved heavy minerals, *Jour. Sed. Petrology* **37**:521-529.

Imbie, J., and T. H. van Andel, 1964, Vector analysis of heavy mineral data, *Geol. Soc. America Bull.* **75**:1131-1156.

Krumbein, W. C., and F. J. Pettijohn, 1938, *Manual of Sedimentary Petrography*, Appleton-Century-Crofts, New York, 320-326, 335, 343, 357-362.

Krynine, P. D., 1942, Provenance versus mineral stability as a controlling factor in the composition of sediments, *Geol. Soc. America Bull.* **53**:1850-1851.

Krynine, P. D., 1946, The tourmaline group in sediments, *Jour. Geology* **54**:65-88.

Lowright, R., E. G. Williams, and F. Dacaille, 1972, An analysis of factors controlling deviations in hydraulic equivalence in some modern sands, *Jour. Sed. Petrology* **42**:635-645.

Marshall, B., 1967, The present status of zircon, *Sedimentology* **9**:119-136.

Marshall, C. E., and C. D. Jeffries, 1945, Mineralogical methods in soil research: I. The correlation of soil types and parent materials with supplementary information on weathering processes, *Soil Sci. Soc. America Proc.* **10**:397-405.

Milner, H. B., 1962, *Sedimentary Petrography*, 2 vols., Allen & Unwin, London, 643p., 715p. (Vol. 1 discusses techniques and vol. 2, specific minerals.)

Parfenoff, A., C. Pomerol, and J. Tourenq, 1970, *Les Mineraux en Grains*, Masson, Paris, 578p.

Pettijohn, F. J., 1941, Persistence of heavy minerals in geologic time, *Jour. Geology* **49**:610-625.

Pryor, W. A., and N. C. Hester, 1969, X-ray diffraction analysis of heavy minerals, *Jour. Sed. Petrology* **39**:1384-1389.

Rice, R. M., D. S. Gorsline, and R. H. Osborne, 1976, Relationships between sand input from rivers and the composition of sands from the beaches of Southern California, *Sedimentology* **23**:689-703.

Rittenhouse, G., 1943, Transportation and deposition of heavy minerals, *Geol. Soc. America Bull.* **54**:1725-1780.

Rosenblum, S., 1958, Magnetic susceptibilities of minerals in the Frantz isodynamic magnetic separator, *Am. Mineralogist* **43**:170-173.

Ross, C. S., 1926, Methods of preparation of sedimentary materials for study, *Econ. Geology* **21**:454-468.

Rubey, W. W., 1933, The size distribution of heavy minerals within a water-laid sandstone, *Jour. Sed. Petrology* **3**:3-29.

Slingerland, R. L., 1977, The effects of entrainment on the hydraulic equivalence relationships of light and heavy minerals in sands, *Jour. Sed. Petrology* **47**:753-776.

Stapor, F. W. Jr., 1973, Heavy mineral concentrating processes and density/shape/size equilibria in the marine and coastal dune sands of the Apalachicola, Florida, region, *Jour. Sed. Petrology* **43**:396-407.

Stow, M. H., 1938, Dating Cretaceous-Eocene tectonic movements in Big Horn Basin by heavy minerals, *Geol. Soc. America Bull.* **69**:731-762.

Tester, A. C., 1932, Abstracts of literature on accessory minerals in sedimentary rocks as related to possible source crystalline rocks, in Report of the Committee on Sedimentation, 1930-1932, *Natl. Research Council Bull.* **89**:168-182.

van Andel, T. J. H., 1959, Reflections on the interpretation of heavy mineral analyses, *Jour. Sed. Petrology* **29**:153-163.

Young, E. J., 1966, A critique of methods for comparing heavy mineral suites, *Jour. Sed. Petrology* **36**:57-65.

Clay Minerals

Brindley, G. W., and G. Brown, 1980, *Crystal Structures and Clay Minerals and Their X-ray Identification*, Mineralogical Society, London, 497p.

Carroll, D., 1959, Ion exchange in clays and other minerals, *Geol. Soc. America Bull.* **70**:749-780.

de Segonzac, G. D., 1970, The transformation of clay minerals during diagenesis and low grade metamorphism: review, *Sedimentology* **15**:281-246.

Fieldes, M., and L. D. Swindale, 1954, Chemical weathering of silicates in soil formation, *New Zealand Jour. Sci. Technology* **36B**:140-154.

Fripiat, J. J., (ed.), 1982, *Advanced Techniques for Clay Mineral Analysis*, Developments in Sedimentology, vol. 34, Elsevier, Amsterdam, 235p.

Gibbs, R. J., 1968, Clay mineral mounting techniques for X-ray diffraction analysis: a discussion, *Jour. Sed. Petrology* **38**:242-244.

Gibbs, R. J., 1977, Clay mineral segregation in the marine environment, *Jour. Sed. Petrology* **47**:237–243.

Grim, R. E., 1968, *Clay Mineralogy*, McGraw-Hill, New York, 384p.

Hayes, J. B., 1970, Polytypism of chlorite in sedimentary rocks, *Clays and Clay Minerals* **18**:205–306.

Keller, W. D., 1970, Environmental aspects of clay minerals, *Jour. Sed. Petrology* **40**:788–813.

MacKenzie, R. C. (ed.), 1957, *The Differential Thermal Analysis of Clays*, Mineralogical Society, London, 456p.

Millot, G., 1970, *Geology of Clays: Weathering, Sedimentology, Geochemistry*, Springer-Verlag, New York, 429p.

Milne, I. H., and J. W. Earley, 1958, Effect of source and environment on clay minerals, *Am. Assoc. Petroleum Geologists Bull.* **42**:328–338.

Mueller, G., 1967, *Methods in Sedimentary Petrology*, H.-W. Schmincke, trans., Hafner, New York, 283p.

Powers, M. C., 1957, Adjustment of land-derived clays to the marine environment, *Jour. Sed. Petrology* **27**:355–372.

Shutov, V. D., A. V. Alexsandrova, and S. A. Losievskaya, 1971, Genetic interpretation of the polymorphism of the Kaolinite group in sedimentary rocks, *Sedimentology* **15**:69–82.

Singer, A., 1980, The paleoclimatic interpretation of clay minerals and weathering profiles, *Earth-Sci. Rev.* **15**:303–326.

Staub, J. R., and A. D. Cohen, 1978, Kaolinite-enrichment beneath coals: a modern analogue, Snuggedy Swamp, South Carolina, *Jour. Sed. Petrology* **48**:203–210.

Syvitski, J. P. M., and A. G. Lewis, 1980, Sediment ingestion by *Tigriopus californicus* and other zooplankton: mineral transformation and sedimentological considerations, *Jour. Sed. Petrology* **50**:869–880.

Warshaw, C. M., and R. Roy, 1961, Classification and a scheme for the identification of layer silicates, *Geol. Soc. America Bull.* **72**:1455–1492.

Weaver, C. E., 1967, The significance of clay minerals in sediments, in B. Naǵy and U. Colombo (eds.), *Fundamental Aspects of Petroleum Geochemistry*, Elsevier, Amsterdam, 37–76.

Weaver, C. E., and L. D. Pollard, 1973, *The Chemistry of Clay Minerals*, Developments in Sedimentology vol. 15, Elsevier, Amsterdam, 213p.

Wilson, M. D., and E. D. Pittman, 1977, Authigenic clays in sandstones: recognition and influence on reservoir properties and paleoenvironmental analysis, *Jour. Sed. Petrology* **47**:3–31.

Wilson, M. J., and D. R. Clark, 1978, X-ray identification of clay minerals in thin section, *Jour. Sed. Petrology* **48**:656–660.

Worrall, W. E., 1968, *Clays: Their Nature, Origin and General Properties*, MacLaren & Sons, London, 128p.

GLAUCONITE

Burst, J. F., 1958, Glauconite pellets: their mineral nature and applications to stratigraphic interpretations, *Am. Assoc. Petroleum Geologists Bull.* **42**:310–327.

Kossovskaya, A. G., and V. R. Drits, 1970, Micaceous minerals in sedimentary rocks, *Sedimentology* **15**:83–101.

McConchie, D. M., and D. W. Lewis, 1978, Authigenic, perigenic, and allogenic glauconites from the Castle Hill Basin, North Canterbury, New Zealand, *New Zealand Jour. Geology and Geophysics* **21**:199–214.

McConchie, D. M., and D. W. Lewis, 1980, Varieties of glauconite in late Cretaceous and early Tertiary rocks of the South Island of New Zealand, and new proposals for classification, *New Zealand Jour. Geology and Geophysics* **23**:413–437.

McConchie, D. M., J. B. Ward, V. H. McCann, and D. W. Lewis, 1979, A Mossbauer investigation of glauconite and its geological significance, *Clays and Clay Minerals* **27**:339–348.

McRae, S. G., 1972, Glauconite, *Earth-Sci. Rev.* **8**:397–440.

Odin, G. S., and A. Matter, 1981, De glauconiarum origine, *Sedimentology* **28**:611–641.

Triplehorn, D. M., 1966, Morphology, internal structure and origin of glauconite pellets, *Sedimentology* **6**:247–266.

Ward, D. M., and D. W. Lewis, 1975, Paleoenvironmental implications of storm-scoured, ichnofossiliferous mid-Tertiary limestones, Waihao District, South Canterbury, New Zealand, *New Zealand Jour. Geology and Geophysics* **18**:881–908.

Sedimentary Carbonates

MINERALOGY

Most carbonate rocks consist of calcite or dolomite; beyond this gross distinction, mineralogy is not generally a useful parameter for a systematic classification designed to assist interpretation of the history of these rocks.

However, other carbonate minerals do occur commonly in modern sediments, and uncommonly in ancient sedimentary rocks. Compositional/crystallographic properties of the most common carbonates are shown in Table 22.

In particular, the unstable polymorphs of $CaCO_3$ —high-magnesium calcite and aragonite—are common in modern sediments (see Table 23). They are

Table 22. Properties of Common Sedimentary Carbonates

Calcite Group: Hexagonal system, rhombohedral-scalenohedral class. Uniaxial negative.

Calcite $CaCO_3$ — H 3 / G 2.7 — Low-Mg calcite—less than 4% $MgCO_3$ in solid solution; the common, stable form.

Magnesite $MgCO_3$ — H 3.5–4.5 / G 3.0–3.5 — High-Mg calcite—more than 4% $MgCO_3$ in solid solution (up to 30% in some algal hardparts, to 43% in echinoderm teeth). Only precipitated by organisms and only stable while organic tissue is present.

Siderite $FeCo_3$ — H 4–4.5 / G 3.9

Rhodochrosite $MnCO_3$ — H 3.5–4 / G 3.2–4.0

Smithsonite $ZnCO_3$ — H 4–5 / G 4.4

Aragonite Group: Orthorhombic system. Biaxial negative.

Aragonite $CaCO_3$ — H 3.5–4 / G 2.95 — Metastable polymorph of calcite (hence rare in pre-Pleistocene rocks); mainly precipitated by organisms (and stable only while organic tissue is present). Rarely contains up to 1.5 mole % of magnesium.

Witherite $BaCO_3$ — H 3.5 / G 4.3

Strontianite $SrCO_3$ — H 3.5–4 / G 3.71

Cerussite $PbCO_3$ — H 3–3.5 / G 6.57

Dolomite Group: Hexagonal system, rhombohedral-rhombohedral class. Uniaxial negative.

Dolomite $Ca,Mg(CO_3)_2$ — H 3.5–4 / G 2.87 — Well-ordered mineral with alternating layers of $MgCO_3$ and $CaCO_3$; more stable than calcite but seems mostly of diagenetic origin in Phanerozoic rocks.

Ankerite $Ca(Mg, Fe, Mn) (CO_3)_2$ — H 3.5–4 / G 2.9–3.1

Kutnohorite $Ca(Mn)(CO_3)_2$ — v. rare

Uncommon but of economic importance are the Monoclinic Prismatic minerals:

Azurite $Cu_3(OH)_2 (CO_3)_2$ — H 3.5–4 / s.g. 3.80 — biaxial negative

Malachite $Cu_2(OH)_2 (CO_3)$ — H 6 / s.g. 4.05 — biaxial positive

Table 23. Mineralogy of Carbonate Sediments

High-Mg Calcite

common in: benthic foraminifera, sponges, alcyonarian corals, echinodermata, decapods, benthic red algae, some bryozoa and brachiopods

also in a few: annelid tubes, cephalopods, ostracods, barnacles

Aragonite

common in: bryozoa, gastropods, bivalves, scleractinian corals, most cephalopods, pteropods, chitons, annelid tubes, benthic green algae, madreporian corals, Palaeozoic stromatoporoids (?), Mesozoic/Cenozoic anthozoa

also in a few: alcyonarian corals, benthic foraminifera, sponges, bryozoa

Low-Mg Calcite:

common in: planktic foraminifera, brachiopods, trilobites, bivalves, ostracods, barnacles, pelagic algae (coccoliths), some stromatoporoids (?), some bryozoa

Sources: after Chave, 1962; Lowenstam in Chilingar et al. 1967; and Scholle, 1978

Note: Skeletal mineralogy may vary between genera of the same order, or species of the same genera; some organisms precipitate several minerals in their hard parts—e.g., aragonite and calcite in some bivalves, gastropods; high-Mg calcite and aragonite in some bryozoa, annelid tubes.

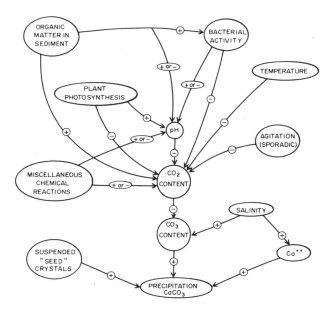

Figure 90. Some factors influencing precipitation of inorganic $CaCO_3$. Where an increase of one increases the other is indicated by + and an antipathetic relationship by −. (*After a sketch by R.L. Folk*)

precipitated mainly by organisms, as are most low-magnesium calcites, but some aragonite and low-magnesium calcite precipitates inorganically (see Fig. 90 for major controls). These unstable carbonates convert to low-magnesium carbonate either in the solid state, via a powder stage, or by dissolving and reprecipitating. They also appear to be easily converted to dolomite in early diagenetic reactions, but low-magnesium calcite also can be transformed in either early or late diagenesis (see Fig. 91 for some major controls).

Dolomite can often be distinguished from calcite and aragonite by its lesser reaction to dilute HCl (only the powder resulting from scratches will react) and by its common tendency to form 1-to 2-mm subhedral-euhedral rhomboids (the rhombic habit of much finer crystals is apparent only with microscope examination, and coarser crystals are commonly anhedral). In thin section, dolomite is almost never twinned (calcite commonly is) and it may show undulose extinction (which calcite almost never does). Staining techniques (see below) or more refined laboratory analysis (such as XRD and DTA) are commonly necessary to distinguish the other carbonate minerals. Siderite is unstable in the normal weathering environment and commonly changes to calcite with exsolution of iron oxides, resulting in a reddish brown coloration that may be the main indication of its original presence.

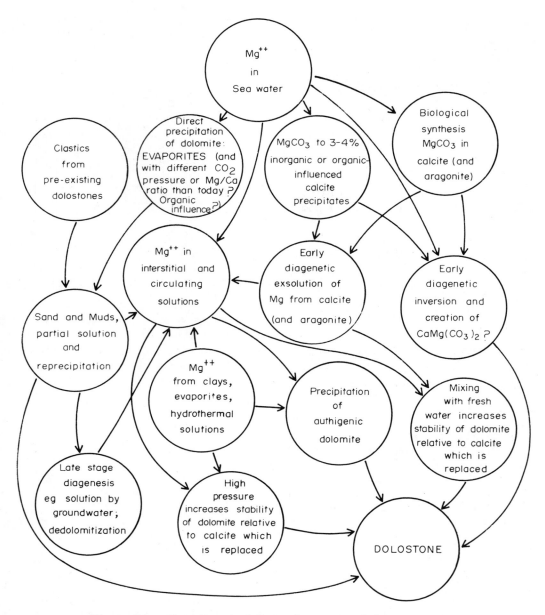

Figure 91.   Genetic possibilities for marine dolostones.

CLASSIFICATION

Composition

Limestone classification systems are numerous (for example, see Ham, 1962). The system most commonly used for composition is that of Folk (1959). It is based on the presence and proportion of the following:

Micrite: lime mud, consisting of microcrystalline (less than 4μ) calcite (originally much may have been aragonite). Generally formed within the environment of sedimentation, this material constitutes the matrix to any grains that may be present. Micrite is the dark "background" in thin sections; the high-power objective is necessary to see individual crystals, which are superposed in a 0.03 mm rock slice. It originates by abrasion of carbonate clasts, disarticulation of skeletal parts of flora and fauna, inorganic precipitation, and diagenetic recrystallization.

Sparite: crystalline carbonate cement comprising clear crystals larger than 10–15μ. It is generally formed after deposition of the surrounding sediment as a void-filling precipitate from supersaturated interstitial waters. However, it may also form through diagenetic or metamorphic recrystallization of micrite or other carbonate grains; evidence for such origin may be crystal boundaries transecting grain outlines, patchy replacement relationships in the specimen, or the presence of *microspar*—1- to 15-μ crystals that generally represent an intermediate stage of growth from micrite to sparry calcite.

Note: The proportion of micrite to sparite is important, inasmuch as lime mud is thought to be forming in every carbonate-accumulating environment; its absence is thus a reflection of current winnowing processes. When calculating the percentage of either micrite or sparite, ignore that which is part of carbonate grains or internal to carbonate grains (these count as allochems).

Allochems: discrete grains, generally transported locally within the environment of their origin. There are four basic types:

1. Fossil fragments: biogenic or "skeletal" components—any size or shape. See Johnson (1951), Majewske (1969), Horowitz and Potter (1971), Bathurst (1975), and Scholle (1978) for photomicrographs of these components as seen in thin section.

2. Pellets: ovoidal/spheroidal, round grains up to 0.2-mm diameter of homogeneous micrite. These are difficult to distinguish, even in thin section. Pellets are rare in Cenozoic rocks. Most originate as fecal ejecta of invertebrates.

3. Oölites: spherical particles up to 2 mm in diameter, showing a concentric or radial structure. The nucleus is another allochem or a detrital grain. Oölites are generally formed in high-energy environments (such as tidal channels) and warm water (supersaturated with $CaCO_3$). (*Superficial oölites* have only one or two concentric coats on the nucleus. *Pisolites* are similar to oölites but larger than 2 mm. Both grain types are formed under different conditions from true oölites).

4. Intraclasts: irregular, rounded, or angular clasts from 0.2 mm to boulder size, consisting of one or more allochems (or detrital grains) generally set in micrite. They may also comprise homogeneous micrite. They represent fragments of sediment from the nearby sea floor, torn up during temporary bursts of high energy while still only partly consolidated. Intraclasts must be distinguished from lithoclasts, which are derived from fully consolidated rocks. Clasts made up of allochems with sparite cement are technically lithoclasts because the source sediment was indurated, but local submarine cementation may be synsedimentary, and some such clasts may be true intraclasts in the Folk sense.

If allochems comprise more than 10% of the rock, it is an allochem rock and the prefix *bio-, pel-, oö-,* or *intra-* is appended to the suffix *-micrite* or *-sparite* (or *-microsparite*), whichever is most abundant. If several allochems are present, several prefixes may be used (in order of abundance)—for example, *biopelsparite.*

If there is a particularly common type of fossil fragment in the rock, append the name of that fossil (at the level you can recognize) to the rock type—for example, *brachiopod-biosparite,* or *Pachymagas-biosparite.*

The classification system appears in Fig. 92.

Figure 92. Classification of carbonate rocks. (After Folk, 1959)

Volumetric Allochem Composition			LIMESTONES, PARTLY DOLOMITIZED LIMESTONES AND PRIMARY DOLOSTONES			Undisturbed Biogenic Rocks	Replacement dolostones
			>10% Allochems		<10% Allochems Microcrystalline Rocks		
			Sparry Cement > Microcrystalline Ooze	Microcrystalline Ooze Matrix > Sparry Cement	1-10% Allochems	<1% Allochems	
					Most abundant Allochem		
>25% Intraclasts			Intrasparite	Intramicrite	Intraclast-bearing Micrite		Coarse, medium, finely, or very finely crystalline dolostone.
<25% Intraclasts	>25% Oölites		Oösparite	Oömicrite	Oölite-bearing Micrite		Micrite or Dolomicrite
	<25% Oölites	Volume ratio of Fossils to Pellets >3:1	Biosparite	Biomicrite	Fossiliferous Micrite		(If "ghosts" of allochems are visible, add their type before "dolostone" - e.g. finely crystalline fossiliferous dolostone)
		3:1 to 1:3	Biopelsparite	Biopelmicrite		Biolithite	
		1:3	Pelsparite	Pelmicrite	Pelletiferous Micrite		

Texture

Limestones with a dominance of sand-size grains are *calcarenites*; those with a dominance of clasts larger than 1 (by Folk) or 2 mm (for consistency with detrital textures) are *calcirudites*; and those with a dominance of lime mud are *calcilutites*. Clast sizes should be specified, but they have much less significance than detrital sediment sizes because the hydrodynamic behavior of skeletal carbonate grains cannot easily be predicted. (Shape and original density are of equal importance to maximum dimension—which often cannot be seen anyway!) For crystal sizes and fabric of the sparite cement and recrystallized carbonate rocks, a separate set of terminology exists (for example, see Folk, 1965), but commonly it is used only where detailed analysis of diagenetic history is attempted.

A special term—*dismicrite*—has been applied by Folk to calcilutites with irregular patches of sparry cement that have infilled voids created by burrowing, gas escape, or other mechanisms that disrupted the initial fabric of the lime mud.

The most common scheme for textural classification is that of Dunham (in Ham, 1962); it is summarized in Table 24. Less easily applied, but worth considering, is Folk's scheme (in Ham, 1962), which would be applied essentially to biolimestones and was devised to parallel the textural maturity scheme for detrital sandstones (see Table 25). (*Note:* Both these schemes are applied only to the *original depositional character* of the sediment; diagenetic modifications must be excluded, or if that is impossible, the schemes should *not* be used.)

Another classification scheme of interest, with an emphasis on texture, is that of Todd (1966), who suggests a triangular diagram on which limestones can be quantitatively plotted.

In all carbonate rocks, any detrital components that may be present can be classified and interpreted as in the detrital rocks (see earlier chapter).

Table 24. Texture of Carbonate Sediments

Depositional Texture

Mud Supported		Grain Supported	
allochems		mud	
<10%	>10%	present	absent
mudstone	***wackestone***	***packstone***	***grainstone***

If depositional texture not recognizable—***crystalline carbonate.*** Subdivide by separate scheme (e.g., see Folk, 1965).

If original components were bound together during deposition—***boundstone*** (for example, parts of reefs, intergrown skeletal components, cavities roofed by organic matter; samples larger than thin section generally required for recognition).

Source: after Dunham in Ham, 1962

Table 25. Folk's Textural Classification Scheme

Percent Allochems	Over ⅔ Lime Mud Matrix				Subequal Spar and Lime Mud	Over ⅔ Spar Cement		
	0-1%	1-10%	10-50%	Over 50%		Sorting Poor	Sorting Good	Rounded and Abraded
Representative Rock Terms	Micrite	Fossiliferous micrite	Sparse biomicrite	Packed biomicrite	Poorly washed biosparite	Unsorted biosparite	Sorted biosparite	Rounded biosparite

171

Example of Refinement in Limestone Classification

To depict quantitative data and sample interrelationships, triangular graphs seem once again most suitable. Any three independent variables may be selected as poles on a triangle, depending on the characteristics of the particular limestone suite and on the worker's own decision as to the most significant parameters for his or her particular purpose.

An example of the kind of triangle that could be used is given in Fig. 93 (constructed in discussion with C. S. Nelson, University of Waikato, Hamilton, New Zealand). Although not as symmetrical as many classifications previously published, the system of subdivision is particularly suitable to temperate-region limestone deposits (such as New Zealand Tertiary varieties: see Nelson, 1978, and Table 26). In these deposits, bioclasts are virtually the only allochems, detrital particles (siliciclasts) are commonly present, and lime mudstones (calcilutites) generally appear to be segregated from cal-

carenites, unlike the case for limestones generated in tropical or subtropical regions (perhaps because of a different principal source for the micrite). For these rocks, logical boundaries between major carbonate-rich rock types are at 50% lime mud (calcilutites versus calcarenites) and 50% siliciclasts (arenites versus calcarenites). Boundaries at 10% of the three major components conveniently differentiate limestones with major fractions of the components from those with minor fractions. The main problem with this system is that it excludes glauconite, a common component of temperate-region limestones. In sequences where glauconite content appears to be a significant variable, either the micrite or siliciclasts pole (whichever appears to be an insignificant variable) could be replaced by glauconite; the term *calcilutite* or *arenite* in the nomenclature given below would be replaced by *greensand*, and *glauconitic* would be substituted for either *sandy* or *micritic*. This scheme is not tested adequately in practice, but it appears viable and serves as an example of the type of refined classification that workers could devise.

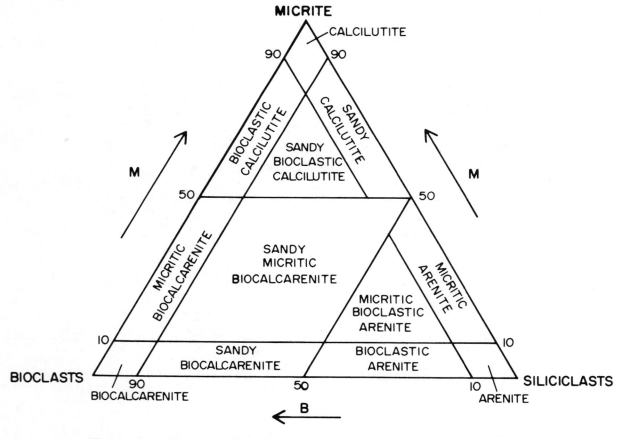

Figure 93. Tentative classification triangle for limestones with siliciclastic components.

172

INTERPRETATIONS

Reefs (biolithites) are very complex and beyond the scope of this brief review. The simplest forms are represented by intercemented bivalves (such as oysters). Reefal deposits must be largely studied in the field; hand specimens commonly cannot be ascribed to such a source. It is worth noting, however, that whereas reefs normally develop in relatively high-energy environments (with much current activity), the reef-building organisms baffle the current action and micrite is common in many deposits.

Most limestones contain abundant fossil debris. Although redeposited, the skeletal elements were generated in the same gross depositional environment as that in which they accumulated and generally have not been transported far from their locus of origin. Hence identification of the organic components is useful to the level (phylum, class, family, genus, even in some cases species) where useful information can be derived from existing knowledge of their paleoecology (see paleontology texts and journals). Even the presence of a single benthic fossil can provide a useful guide to paleo-environmental conditions (for example, the presence of a single oyster in a lime mudstone may indicate the existence of a local firm substrate). When interpreting the original living community in any paleoenvironment, keep in mind not only potential bias resulting from probable selective sorting of debris by processes in the environment and possible selective destruction of remains by scavengers, but also the likelihood that more debris was contributed from short-lived organisms than long-lived ones (several generations of bryozoa, for example, may grow and die during the lifespan of one brachiopod). Burrowing activities of infauna or storm redeposition may also intermix populations that lived in different subenvironments. In some submarine environments, carbonate particles including fossils may be recycled (see Mac-Intyre, 1970).

Fossil constituents also provide an indication of the original mineralogical composition of carbonate sediments (see Table 23, and Majewske, 1969, or Scholle, 1978, for longer tables listing shell mineralogies). Remember, however, that high-magnesium calcite or aragonite components may have been selectively destroyed during their inversion to low-magnesium calcite and that we know little about the original mineralogy of some extinct organisms. Knowledge of the original mineralogy is important in interpretations of diagenetic modifications.

Pellets appear to be rare in Cenozoic relative to Paleozoic deposits; inferences from their presence or absence must thus be restricted to studies of Paleozoic sequences. Oölites appear to require warm (supersaturated) waters. Intraclasts may not be formed even in situations where intermittent high-energy events occur if the cohesion of earlier carbonate sediments has not reached a particular critical state.

For textural interpretations, you must consider not only availability of particular allochems, but their inherent size, shape, and internal structure—some fossil fragments (such as from barnacles or *Inoceramus*) can never become rounded however much abrasion they suffer, whereas oölites and pellets are always well rounded because of their mode of origin. Limestones with sparry cements generally reflect deposition in environments with consistent winnowing activity, such as dunes, offshore bars, beaches, and other current-swept settings (exception: where the sparry cement results from diagenetic recrystallization of micrite). However, presence or absence of micrite must be evaluated in the full understanding not only of the multiple modes of origin of this sediment (for example, see Matthews, 1966; Bathurst, 1970; Stockman et al., 1967; Alexandersson, 1979; Stieglitz, 1973; Perkins and Halsey, 1971) but also of the potential exceptions to the assumption that energetic currents will remove it (marine vegetation may baffle the currents, algae may trap or bind the mud, irregular shells may produce local low-energy pockets, micrite may be introduced or formed after final deposition).

Dolomite, while generally rare in modern sediments, is next in abundance to calcite in ancient carbonate rocks. Genetic possibilities for dolomite are outlined in Fig. 91; the consensus is that it generally results from diagenetic replacement of calcite and aragonite. However, it is difficult to interpret the origin of some thick, continuous, non porous, fine dolostones, and there are various settings in which diagenetic dolomite can form (see Shinn et al., 1965; Zenger, 1972; Badiozamani, 1973; Folk and Land, 1975; and textbooks).

Fig. 94 and Tables 26 and 27 illustrate salient aspects of carbonate environments. Consult Bathurst (1975) for a comprehensive survey of modern tropical/subtropical carbonate environments and deposits. See Roehl (1967) and Eliuk (1978) for interesting examples of carbonate studies, and Williamson and Picard (1974) for an example of the complex carbonate facies that may develop in nonmarine settings.

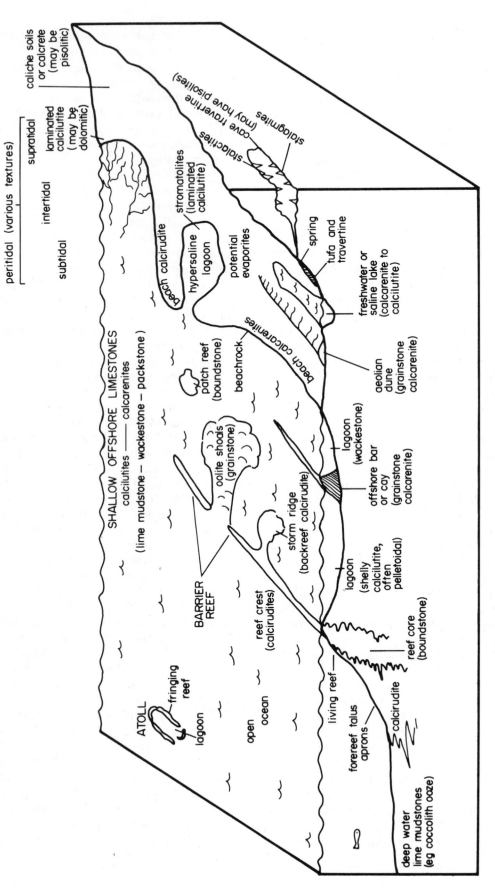

Figure 94. Schematic diagram of potential carbonate environments.

Table 26. Comparison of Some Tropic-Subtropic and Temperate Shelf Carbonate Parameters

	Tropical-Subtropical	Temperate Shelf Carbonates (New Zealand Cenozoic)
Latitude	Between 30°N and 30°S	Between 60°S and 35°S
Mean annual water temperature	>23°C	<20°C
Minimum water temperature	14°C	About 5°C
Water salinity	30 to >50%	30–35%
Level of CaCO₃ saturation	Supersaturation	Undersaturation to supersaturation
Shelf gradient	<0.5 m/km	>0.5 m/km
Reef structures	Common (mainly coralgal)	Rare (mainly oyster)
Sedimentation rate	10–100 cm/1000 yr	<5 cm/1000 yr
CaCO₃ content	>90%	50–100%
Terrigenous grains	Rare	Rare to abundant
Glauconite	Rare	Common
Evaporite minerals (including dolomite)	Common	Absent
Nonskeletal carbonate grains	Common to abundant	Rare (mainly intraclasts; no oolites)
Major skeletal grain types	Calcareous green algae Corals (hermatypic) Benthic foraminifera Molluscs Calcareous red algae	Bryozoans Echinoderms Benthic foraminifera Calcareous red algae Barnacles Molluscs Brachiopods
Algal mats	Common	Absent
Carbonate mud	Common to abundant	Absent to common
Primary sediment mineralogy	Aragonite > Mg-calcite	Mg-calcite >> calcite > aragonite
Environment of alteration of metastable carbonate grains	Subaerial	Submarine
Environment of major lithification	Submarine and subaerial	Shallow subsurface
Mineralogy of major cement	Aragonite, Mg-calcite, or calcite	Calcite
Major sources of cement	Sea water and solution of aragonite grains	Selective and nonselective intergranular solution of calcite skeletons

Source: after Nelson, 1978

Table 27. Some Characteristics of Platform Carbonate Sediments

Morphologic Zones:	>100 km Open Sea	10–100 km? Barrier, Reef, or Shelf Edge	0 to >100 km Lagoon or Shelf Interior	<10 km Intertidal	<15 km Supratidal
Energy Sources:	Oceanic Currents	Waves and Tides	Waves, Currents, and Storms . . . . Tides . . . . Wind		
Texture:	Muds; Sandy Muds	Sands; Gravels	(Sandy) Muds	Variable	
SEDIMENT STRUCTURES					
Lamination	abundant	rare	. . . . abundant		
Cross-bedding		abundant		. . . . possible	
Dessication structures					common . . . . . .
Reefs		common	patch		
Burrows	common		abundant	common	
Stromatolites				. . . . common	
SEDIMENTS					
Skeletal sands	abundant	abundant	abundant	common	
Ooids		abundant			
Pellets		. . . . . abundant			
Aggregates		. . . . . abundant			
Intraclasts				. . . . common	
Lime Mud	pelagic; skeletal abrasion	. . . . skeletal abrasion . . . . ; . . . . floral disarticulation			inorganic
Evaporites				. . . . common	
BIOTIC ELEMENTS					
Marine grasses		common	abundant	common	
Blue-green algae					abundant
Green algae		common	abundant		
Red algae		abundant	common		
Pelagic algae	abundant		common		
Corals		. . . . abundant	possible		
Molluscs	present	common	abundant	present	
Planktic forams	abundant	common	sparse–common		
Benthic forams	common	common	abundant	present	

Source: after Nelson, 1978

METHODS OF STUDY

The texture of limestone hand specimens is clarified by moistening the surface with a clear oil (or water). Etching with dilute HCl results in differential relief of components, especially dolomite and other insoluble residues. Acetate peels (see below) are easily prepared if smooth surfaces are available or can be made (for example, with sandpaper); detail is as good as obtained from thin section and broader areas can be examined. An acidic solution of alizarin-red may be applied in the field as well as the laboratory to distinguish calcite and aragonite from dolomite (this and other staining procedures for more refined mineralogical identification are described below). Complete digestion of samples in acetic acid or HCl permits study of any insoluble components, although some minerals such as the clays may be affected by the acid.

Thin Section Studies

Fig. 95 provides an example of a worksheet for describing limestones in thin section; depending on the characteristics of your particular suite of carbonate rocks, you may wish to design a different format.

Acetate Peels for Studying Limestone Textures

Acetate (celluloid) peels are used primarily to study limestone textures. Etching in acid of polished rock surfaces produces a differential relief because constituents dissolve at different rates depending on their density, crystal size, and crystal orientation. The peels mold themselves to the irregularities and form castings that can be studied with a microscope and/or used to prepare photographic contact prints (use a high-contrast paper, such as Kodabromide F-5, and set the lens at approximately f/11 for a good depth of field). The method has several advantages over thin-section studies: analysis is easier and more rapid; the cost is lower; greater surface area is available for study; and details are generally seen as well as, or better than, in thin section. Details of peel preparation vary; experimentation is necessary for particular rock types. Generally, the procedure is as follows:

(Peel techniques are also applicable to broad surfaces of friable sediment [any composition] and even to silicate rocks, for which the technique is as described below except that etching is done with HF fumes or even 5–40% HF solutions—*be careful!* A comprehensive discussion of peel techniques is provided by Bouma, 1969.)

1. Polish the surface of the sample (with sandpaper and emery paper or carborundum powder down to 800 grade, depending on the rock and the desired degree of detail).

2. Etch the surface uniformly for 10–45 sec with 5, 10, or 15% HCl. (The easiest method is to dip the surface into a bowl of acid.) Avoid excessive effervescence, which will result in uneven etching.

3. If desired, apply stains for mineralogical discrimination (see below).

4. Wash gently with distilled water, and dry thoroughly.

5. Arrange the etched surface to face upward and to lie horizontally (embed the base in a bowl of sand).

6. Cover the surface with acetone, allow excess fluid to evaporate, and then while the entire surface is glistening, apply acetate sheet (medium thickness is usually best). The sheet should cover the entire surface. It is best applied by pinching the edges inward to form a U-shape, then applying the base of the U to the center of the surface and flattening it out from the center to eliminate air bubbles. The matte surface of the acetate sheet should be down. Applying an even weight (such as a sandbag) over the peel for about 10 sec helps ensure even contact.

7. Allow the peel to dry fully (this may require more than 1 hr). Remove the peel carefully. If it sticks, the peel is probably not yet dry.

8. Trim the acetate sheet and mount it onto glass (either by sandwiching it between glass plates or by sealing the edges of the sheet to glass using a brush soaked in acetone).

Staining for Carbonate Mineralogy

There are many staining methods for identifying various carbonate minerals and cations in carbonates (see Wolf, Easton, and Warne, 1957, in Chilingar et al., 1967). Twelve carbonates are economically important (the greatest variety appear to occur in coal sequences, where a great range of microchemical environmental conditions may be

Sample No: Analyst:
Formation: Date:
Location: Age:

Hand Specimen

Colour: Induration:
Porosity/Permeability: Structures:
Max size clast:
Modal Size: Calcirudite/Calcarenite/Calcilutite
Dunham Texture: Folk Texture:

Composition:
Other Features:

THIN SECTION/ACETATE PEEL

        Grains   Size   Roundness   Variety   Percent
Allochems: Bioclasts

           Intraclasts
           Oolites
           Pellets
Detritals:

Perigenic non-
      carbonate
Authigenic non-
      carbonate
Micrite   %
Microsparite   %
Sparite   %

Other Features:

Classification:

                                              (calcirudite)
colour-induration-sed. structures-sorting-size- (calcarenite) - Dunham texture:
                                              (calcilutite)

detrital component-common perigenic/authigenic non-carbonates-Folk compositional term.

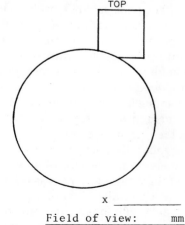

Field of view: mm

Figure 95.   Example of a worksheet for limestone petrography.

178

represented), but only a few tests for the most common carbonate varieties are given below.

Several stain tests are applicable in the field, but for detailed studies a polished face is necessary. A stain solution may generally be used only once, so be careful of quantities used for one sample. Use distilled water for rinsing.

1. Polish a surface of the sample with 400 to 800 carborundum powder, depending on the rock texture and detail desired. Alternatively, uncover a thin section or mount loose grains on a glass slide.

2. Etch the polished surface with approximately 10% HCl. If a brisk reaction occurs, allow 30 sec or less; if little or very little reaction occurs, etch for 45 sec in warm 10% HCl. (If you are using a thin section, etch it in cold acid for 10–15 sec with approximately 1.5% HCl.)

3. Apply stains according to the flowsheet illustrated in Fig. 96; details for the stains used, as well as several others that may prove useful in particular cases, are given below. See also Friedman (1959), Warne (1962), and Dickson (1966).

STAINING WITH ALIZARIN-RED S

1. Use 0.1 g sodium alizarin sulfonate per 100 ml 0.2% cold HCl. (or 2 ml commercial HCl in 998 ml distilled water). For thin sections, use 0.2 g/100 ml.

2. Cover the etched surface with the solution for approximately 5 min (30–45 sec for thin sections).

3. Rinse gently but swiftly, as the dye is soluble.

Deep red indicates calcite, witherite, high-magnesium calcite, or aragonite.
Purple indicates ankerite, ferrodolomite, strontianite, or cerussite.
If the carbonate will not take the stain, it is siderite, dolomite, rhodochrosite, magnesite, or smithsonite, (or it may be anhydrite, or gypsum).

Varieties of unstained or purple minerals may be distinguished by boiling them in a mixture of the stain and 30% NaOH for approximately 5 min (see Fig. 73). (This test also distinguishes high-magnesium calcite from ankerite and calcite.)

FEIGL'S SOLUTION

1. Add 1 g Ag_2SO_4 to a solution of 11.8 g $MnSO_4\cdot 7H_2O$ in 100 ml distilled water. Boil.

2. After cooling, add 1 or 2 drops of dilute NaOH solution. Filter off the precipitate after 1 or 2 hr.

3. Store the solution in a dark bottle.

4. After etching the sample, immerse it in cold Feigl's solution for approximately 10 min.

Aragonite will turn black, and calcite will be unaffected.

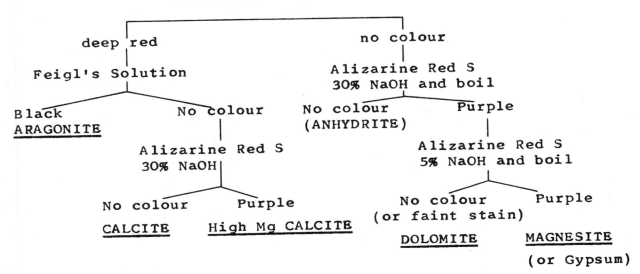

Figure 96. A suggested procedure for staining carbonates. *(After Friedman, 1959)*

179

STAINING WITH POTASSIUM FERRICYANIDE

This stain works only for carbonates that react with the acid medium.

1. Mix equal parts of 2% HCl and K-ferricyanide (2 g/100 ml distilled water). **Caution:** This solution is unstable and gives off poisonous HCN. Keep it under a hood.

2. Wash the sample gently but swiftly. Carbonates with Fe^{2+} are stained blue:
Ankerite and ferrodolomite are dark blue in cold
  solution
Dolomites generally, and siderite always, are stained
  only with heating (up to 5 min for dolomite)
Calcite and magnesite stain if they have any Fe^{2+}.

This solution may be mixed with alizarin-red S (ARS) for a combined test. For thin sections, use three parts ARS to two parts potassium ferricyanide. Calcite should turn pale pink to red, depending on optical orientation. Ferroan-calcite should be the same or a pale to dark blue or mauve, purple, or royal blue. Dolomite should have no color and ferroan-dolomite should be pale to deep turquoise. If you then stain with 0.2 g ARS/100 ml of 1.5% HCl for 10-15 seconds, calcite and ferroan-calcite will turn very pale pink to red and ferroan-dolomite will become colorless.

3. Rinse gently; do not touch the stained surface.

MEIGEN'S SOLUTION (MODIFIED)

1. Prepare a 5-10% solution of cobalt nitrate by mixing 15 g $Co(NO_3)_2$ in 100 ml distilled water.

2. Immerse the sample (not a thin section!) and boil 1-5 min (depending on the grain size).

Aragonite will turn dark violet.

Dolomite, ankerite, magnesite, and siderite will
  remain unstained.
Calcite will remain unstained if coarsely crystalline,
  and will turn lilac-rose if microcrystalline or
  light blue if boiled for a prolonged period.

Boiling time may be critical.

LITERATURE

Alexandersson, E. T., 1979, Marine maceration of skeletal carbonates in the Skagerrak, North Sea, *Sedimentology* **26**:845-852.

Badiozamani, K., 1973, The Dorag dolomitization model-application to the Middle Ordovician of Wisconsin, *Jour. Sed. Petrology* **43**:965-984.

Bathurst, R. G. C., 1970, Problems of lithification in carbonate rocks, *Geol. Assoc. Proc.* **81**:429-440.

Bathurst, R. G. C., 1975, *Carbonate Sediments and Their Diagenesis*, 2nd ed., Developments in Sedimentology 12, Elsevier, Amsterdam, 658p.

Bouma, A. A., 1969, *Methods for the Study of Sedimentary Structures*, Wiley, New York, 458p.

Carozzi, A. V., and D. A. Textoris, 1967, *Paleozoic Carbonate Microfacies of the Eastern Stable Interior (U.S.A.)*, E. J. Brill, Leiden, Netherlands, 146p.

Chave, K. E., 1962, Factors influencing the mineralogy of carbonate sediments, *Limnology and Oceanography* **7**:218-223.

Chilingar, G. V., H. J. Bissell, and R. W. Fairbridge, (eds.), 1967, *Carbonate Rocks, Developments in Sedimentology*, vols. 9A and 9B, Elsevier, Amsterdam, 471p., 413p.

Cloud, P. E. Jr., 1962, Behaviour of calcium carbonate in sea water, *Geochim. et Cosmochim. Acta* **26**:867-884.

Cook, H. E., and P. Enos (eds.), 1977, *Deep-Water Carbonate Environments*, SEPM Spec. Pub. No. 25, Society of Economic Paleontologists and Mineralogists, Tulsa, Okla., 336p.

Davies, P. J., B. Bubela, and J. Ferguson, 1978, The formation of oöids, *Sedimentology* **25**:703-730.

Dickson, J. A. D., 1966, Carbonate identification and genesis as revealed by staining, *Jour. Sed. Petrology* **36**:491-505.

Donahue, J., 1969, Genesis of oölite and pisolite grains: an energy index, *Jour. Sed. Petrology* **39**:1399-1411.

Eliuk, L. S., 1978, The Abenaki Formation, Nova Scotia Shelf, Canada—a depositional and diagenetic model for a Mesozoic carbonate platform, *Canadian Petroleum Geologists Bull.* **26**:424-514.

Folk, R. L., 1959, The practical petrographical classification of limestones, *Am. Assoc. Petroleum Geologists Bull.* **43**:1-38.

Folk, R. L., 1962, Spectral subdivision of limestone types, *Am. Assoc. Petroleum Geologists Mem.* **1**:62-84.

Folk, R. L., 1965, *Some Aspects of Recrystallization in Ancient Limestones*, SEPM Spec. Pub. No. 13, Society of Economic Paleontologists and Mineralogists, Tulsa, Okla., 14-48.

Folk, R. L., 1973, Carbonate petrography in the post-Sorbian age, in R. N. Ginsburg (ed.). *Evolving Concepts in Sedimentology*, John Hopkins University Press, Baltimore, 118-158.

Folk, R. L., and L. S. Land, 1975, Mg/Ca ratio and salinity: two controls over crystallization of dolomite, *Am. Assoc. Petroleum Geologists Bull.* **59**:60-68.

Friedman, G. M., 1959, Identification of carbonate minerals by staining methods, *Jour. Sed. Petrology* **29**:87-97.

Friedman, G. M. (ed.), 1969, *Depositional Environments in Carbonate Rocks*, SEPM Spec. Pub. No. 14, Society of Economic Paleontologists and Mineralogists, Tulsa, Okla., 209p.

Friedman, G. M. (ed.), 1974, Comparative sedimentology of carbonates symposium, *Am. Assoc. Petroleum Geologists Bull.* **58**:781-867.

Friedman, G. M., 1975, The making and unmaking of limestones or the downs and ups of porosity, *Jour. Sed. Petrology* **45**:379-398.

Ham, E. W. (ed.), 1962, *Classification of Carbonate Rocks: A Symposium*, AAPG Mem. 1, American Association of Petroleum Geologists, Tulsa, Okla., 279p.

Horowitz, A. S., and P. E. Potter, 1971, *Introductory Petrography of Fossils*, Springer-Verlag, New York, 300p.

Ingerson, E., 1962, Problems of the geochemistry of sedimentary carbonate rocks, *Geochim. et Cosmochim. Acta* **26**:815-847.

Johnson, J. H., 1951 (1971), An introduction to the study of organic limestones, *Colorado School Mines Quart.* 66, 185p.

MacIntyre, I. G., 1980, Sediments off the West Coast of Barbados: diversity of origins, *Marine Geology* **9**:5-23.

Majewske, O. P., 1969, *Recognition of Invertebrate Fossil Fragments in Rocks and Thin Sections*, E. J. Brill, Leiden, Netherlands, 101p. (plus 106 plates).

Matthews, R. K., 1966, Genesis of lime and mud in Southern British Honduras, *Jour. Sed. Petrology* **36**:428-554.

Perkins, R. D., and S. D. Halsey, 1971, Geologic significance of microboring fungi and algae in Carolina shelf sediments, *Jour. Sed. Petrology* **41**:843-853.

Nelson, C. S., 1978, Temperate shelf carbonate sediments in the Cenozoic of New Zealand, *Sedimentology* **25**:737-771.

Roehl, P. O., 1967, Stony Mountain (Ordovician) and Interlake (Silurian) facies analogues of recent low-energy marine and sub-aerial carbonates, Bahamas, *Am. Assoc. Petroleum Geologists Bull.* **51**:1979-2032.

Scholle, P. A., 1978, *A Color Illustrated guide to Carbonate Rock Constituents, Textures, Cements and Porosities*, AAPG Mem. 27, American Association of Petroleum Geologists, Tulsa, Okla., 241p.

Schwartz, H.-H., 1975, Sedimentary structures and facies analysis of shallow marine carbonates, *Contr. Sedimentology* **3**:1-100.

Sellwood, B. N., 1978, Shallow water carbonate environments, in H. G. Reading (ed.), *Sedimentary Environments and Facies*, Blackwell Scientific Publications, Boston, 259-313.

Shearman, D. J., J. Twyman, and M. Z. Karimi, 1970, The genesis and diagenesis of oölites, *Geol. Assoc. Proc.* **81**:561-575.

Shinn, E. A., R. N. Ginsburg, and R. M. Lloyd, 1965, *Recent Supratidal Dolomite from Andros Island, Bahamas*, SEPM Spec. Pub. No. 13, Society of Economic Paleontologists and Mineralogists, Tulsa, Okla., 112-123.

Stieglitz, R. D., 1973, Carbonate needles: additional organic sources, *Geol. Soc. America Bull.* **84**:927-930.

Stockman, K. W., R. N. Ginsburg, and E. A. Shinn, 1967, The production of lime mud by algae in South Florida, *Jour. Sed. Petrology* **37**:633-648.

Todd, T. W., 1966, Petrogenetic classification of carbonate rocks, *Jour. Sed. Petrology* **36**:317-340.

Warne, S. St. J., 1962, A quick field or laboratory staining scheme for the differentiation of the major carbonate minerals, *Jour. Sed. Petrology* **32**:29-38.

Williamson, C. R., and M. D. Picard, 1974, Petrology of carbonate rocks of the Green River Formation (Eocene), *Jour. Sed. Petrology* **44**:738-759.

Wilson, J. L., 1975, *Carbonate Facies in Geologic History*, Springer-Verlag, New York, 471p.

Zenger, D. H., 1972, Significance of supratidal dolomitization in the geologic record, *Geol. Soc. America Bull.* **83**:1-12.

Zenger, D. H., J. B. Dunham, and R. L. Ethington (eds.), 1980, *Concepts and Models of Dolomitization*, SEPM Spec. Pub. No. 28, Society of Economic Mineralogists and Paleontologists, Tulsa, Okla., 320p.

Evaporite Deposits

Evaporite minerals form wherever evaporation is equal to or greater than the sum of precipitation, runoff, and influx from seawater. Natural evaporite deposition occurs from lakes, springs, groundwater and seawater (see Krumbein, 1951; Scruton, 1953; Stewart, 1963; Borchert and Muir, 1964). Marine evaporites are most common and widespread in the geological column, occurring either as relatively thin and impure deposits of mainly calcium sulfate that accumulated in peritidal paleoenvironments (such as sahbkas), or as thick (to over 1000 m), widespread (to thousands of km²) basinal sequences that include monomineralic beds of sulfates and more soluble salts (such as the Permian Zechstein sequence of northern Europe).

The most common evaporite minerals are listed in Table 28 (see also Stewart, 1963; Dean and Schreiber, 1978). Fig. 97 presents data relevant to evaporite constituents, models of evaporite deposition, and an example of a known sequence. Calcium carbonate will precipitate after about 50% of an original volume of seawater has evaporated (in nature, however, most may be lost earlier to the hard parts of organisms). Calcium sulfates precipitate after 80% evaporation, halite after about 90%, and the most soluble potassium and mixed potassium/sodium/magnesium salts only after 98% evaporation. Natural precipitation sequences are so complex, though, that it is difficult to account for the existence of thick or extensive monomineralic deposits of any kind more soluble than calcium sulfate. Diagenetic (more appropriately, "metamorphic") modifications to original textures and composition are generally substantial.

For the relatively thin, impure, sulfate-dominated evaporite deposits, the model generally envisioned (and documented in modern settings—see Kinsman, 1976; Arakel, 1981) is a marginal marine setting where saline water, derived from saline ground-

Table 28. Common Evaporite Minerals

Sulfates

Anhydrite	$CaSO_4$
Gypsum	$CaSO_4 \cdot 2H_2O$
Kieserite	$MgSO_4 \cdot H_2O$
Langbeinite	$K_2Mg_2(SO_4)_3$
Kainite	$KMg(SO_4)Cl \cdot 3H_2O$
Polyhalite	$CaSO_4 \cdot MgSO_4 \cdot K_2SO_4 \cdot 2H_2O$
(Many others)	

Sulfur—secondary products of sulfates

Borates

Boracite	$Mg(Fe, Mn)_3B_7O_{13}Cl$
Borax	$Na_2B_4O_7$

Carbonates

Aragonite	$CaCO_3$
Dolomite	$Ca, Mg(CO_3)_2$
Magnesite	$MgCO_3$

Chlorides (Many other complex ones)

Halite	$NaCl$
Sylvite	KCl
Carnallite	$KMgCl_3 \cdot 6H_2O$

Nitrates

Complex generally; simplest is

Saltpeter	$NaNO_3$

A

COMPOUND	CONCENTRATION IN SEAWATER (ppt)	Volume H$_2$O	LIMITS FOR PRECIPITATION				DEPTH OF MARINE WATER (M) EVAPORATED FOR 1M PRECIPITATE
			Lower Limit Density	Salinity	Upper Limit Density	Salinity	
CaCO$_3$	0.12	53-19%	1.0500	72	1.264	200	25,000
CaSO$_4$($\cdot$2H$_2$O)	1.27	19-3%	1.1264	200	1.2570	427	2,100
NaCl	27.2	9.5-1.6%	1.2138	353	?		73
Mg SO$_4$	2.25	9.5-0%	1.2138	353	?		
Mg Cl$_2$	3.35	9.5-0%	1.2138	252	?		
K Cl	0.74	1.5-0%	?		?		

N.B. Precipitation in later stages of evaporation is complex and depends on T$^{\circ}$C and other factors

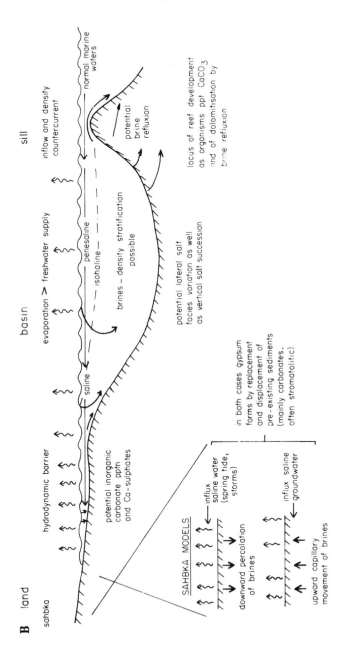

Figure 97. Selected data and models relevant to evaporite deposition.
A: Guideline data on precipitation of some common salts. *(After Stewart, 1963; Dean et al., 1978; Rosler and Lange, 1972)* B: An idealized model for evaporite settings. *(Continues on next page.)*

C

Germany (normal succession)			Whitby District England	
Thickness (m)	Interval		Interval	Thickness (m)
5			Upper Permian Marl	up to 183
117	Fourth Evaporite		top anhydrite	1
15			salt clay	2-4
160-271	Third Evaporite	Upper evaporites	Upper halite potash salts lower halite anhydrite carbonate	15-27 0-9 13-20 5-9 1
8	(dolomitic top)		carnallitic marl	9-19
49-116	Second Evaporite	middle evaporites	upper halite potash salts lower halite halite-anhydrite anhydrite	0-5 3-4 28-84 15-28
9	dolomite and other rocks		Upper Magnesian Limestone	36-56
57	First evaporite	lower evaporites	upper halite-anhydrite upper anydrite lower halite-anhydrite lower anydrite	12-36 12-183 14-43 46-93
6	chalk Kupferschiefer (sulphides) conglomerate		Lower Magnesian Limestone (some anhydrite) basal sands, breccias, marls	111

N.B. thicknesses vary locally; correlations approximate; subdivisions not given for German succession.

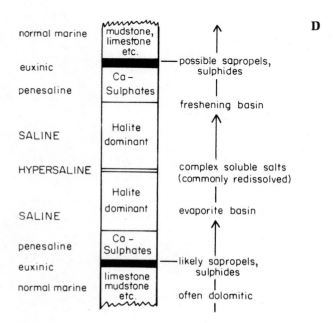

D

Figure 97. *(Continued)* C: Zechstein basin sequence. *(After Stewart, 1963)* D: Ideal evaporite cycle.

water or from sporadic seawater flooding, evaporates near or at the surface with concommittant precipitation of its least soluble salts into the pores of the sediment. Later (diagenetic) reactions may result in replacement or displacement of primary clastic sediments (such as limestones). Displacement by growing sulfate nodules often produces "chicken-wire" structure, where the primary sediment is compressed between the evaporites. Coastal salt pans also produce thin but often relatively pure sulfate deposits (for example, see Kushnir, 1981).

For the thick, extensive, monomineralic deposits, the general model imagined is a largely enclosed basin with a restricted access route (the *sill*, as in door sill) to the open sea. In this setting, influx of seawater may balance evaporation at any stage such that precipitation of only one salt occurs. An "ideal" evaporite cycle of deposits from such a basin would begin with normal open marine sediment, gradually progress to the hypersaline stage, then "freshen" to open marine sediments again (Fig. 97). Intermediate stages in basin restriction establish conditions suitable for accumulation of petroleum source materials (and later impermeable salt deposits form caprocks) or for sulfides (for example, the Kupferschiefer deposits in the Zechstein deposits). (See Brongersma-Sanders, 1971, for a general discussion; and see Degens and Ross, 1969, and Friedman 1972, for accumulations in the modern Red Sea.) Many unresolved problems remain to account for the "saline giants," for which we have no good modern analogue (the Miocene evaporites in the Mediterranean basin are the youngest such deposits—see Schreiber et al., 1971; Hsu, 1972). Both deep-water (for example, Schmalz, 1969) and shallow-water (for example, Hardie and Eugster, 1971) models have been proposed, and the complex role of brine stratification and interaction (see Raup, 1970) is still not fully understood.

Because of the solubility of evaporites, it is likely that many original deposits have been eliminated in subsequent geological history (for example, even in the most arid regions, groundwater dissolution prevents exposure of salts more soluble than the sulfates). Clues attesting their original presence may be difficult to find (for example, see Folk and Pittman, 1971) or may occur in the form of collapse features in younger sequences.

A particularly intriguing problem is to account for the enormous deposits of Permian evaporites around the world (the Perm of USSR, the Zechstein Basin of northern Europe, the Midland-Delaware Basin of the southwestern USA, and in Saskatchewan, Canada). It is estimated that the evaporites in the Zechstein Basin alone total about 2×10^9 km^3; in contrast, the total volume of salt in the oceans (of similar salinity throughout the Phanerozoic, by most evidence) is estimated at just over 2×10^{10} km^3 (see Borchert and Muir, 1964). These occurrences are not easily related to Permian glaciation, particularly since Pleistocene glaciation appears to have been more widespread and yet extensive evaporite deposition was not associated with it.

LITERATURE

Anderson, R. Y., W. E. Dean, Jr., D. W. Kirkland, and H. I. Snider, 1972, Permian Castile varved evaporite sequence, West Texas and New Mexico, *Geol. Soc. America Bull.* **83**:59-86.

Arakel, A.V., 1981, Coastal sabkha and salt pan deposition in Hutt and Leeman Lagoons, Western Australia, *Jour. Sed. Petrology* **50**:1305-1326.

Borchet, T. H., and R. O. Muir, 1964, *Salt Deposits: The Origin, Metamorphism, and Deformation of Evaporites*, D. van Nostrand, New York, 338p.

Briggs, L. I., 1958, Evaporitic facies, *Jour. Sed. Petrology* **28**:46-56.

Brongersma-Sanders, M., 1968, On the geographical association of stratabound ore deposits with evaporites, *Mineralium Deposita* **3**:286-291.

Brongersma-Sanders, M., 1971, Origin of major cyclicity of evaporites and bituminous rocks: an actualistic model, *Marine Geology* **11**:123-144.

Dean, W. E., and B. C. Schreiber, 1978, *Marine Evaporites*, SEPM Short Course No. 4, Society of Economic Paleontologists and Mineralogists, Tulsa, Okla., 188p.

Degens, E. T., and D. A. Ross, 1969, *Hot Brines and Recent Heavy Metal Deposits in the Red Sea*, Springer-Verlag, Berlin, 600p.

Folk, R. L., and J. S. Pittman, 1971, Length-slow chalcedony: a new testament for vanished evaporites, *Jour. Sed. Petrology* **41**:1045-1058.

Folk, R. L., and A. Siedlecka, 1974, The schizohaline environment: its sedimentary and diagenetic fabrics as exemplified by Late Paleozoic rocks of Bear Island, Svalbard, *Sed. Geology* **11**:1-15.

Friedman, G. M., 1972, Significance of Red Sea in problem of evaporites and basinal limestones, *Am. Assoc. Petroleum Geologists Bull.* **56**:1072-1086.

Hardie, C. A., and H. P. Eugster, 1971, The depositional environment of marine evaporites: a case for shallow clastic accumulation, *Sedimentology* **16**:187-220.

Hsu, K. J., 1972, Origin of saline giants: a critical review after the discovery of the Mediterranean Evaporite, *Earth-Sci. Rev.* **8**:371-396.

Kinsman, D. J. J., 1976, Evaporites: relative humidity control of primary mineral facies, *Jour. Sed. Petrology* **46**:273-299.

Kirkland, D. W., and R. Evans, (eds.), 1973, *Marine Evaporites: Origin, Diagenesis and Geochemistry*, Benchmark Papers in Geology, vol. 7, Dowden, Hutchinson & Ross, Stroudsburg, Pa., 426p.

Krumbein, W. C., 1951, Occurrence and lithologic association of evaporites, *Jour. Sed. Petrology* **21**:63–81.

Kushnir, J., 1981, Formation and early diagenesis of varved evaporitic sediments in a coastal hypersaline pool, *Jour. Sed. Petrology* **51**:1193–1203.

Raup, O. B., 1970, Brine mixing: an additional mechanism for formation of basin evaporites, *Am. Assoc. Petroleum Geologists Bull.* **54**:2246–2259.

Rosler, H. J., and H. Lange, 1972, *Geochemical Tables*, Elsevier, Amsterdam, 468p.

Schmalz, R. F., 1969, Deep water evaporite deposition: a genetic model, *Am. Assoc. Petroleum Geologists Bull.* **53**:798–823.

Schreiber, B. C., G. M. Friedman, A. Decima, and E. Schreiber, 1971, Depositional environments of Upper Miocene (Messinian) evaporite deposits of the Sicilian Basin, *Sedimentology* **23**:729–760.

Scruton, P. C., 1953, Deposition of evaporites, *Am. Assoc. Petroleum Geologists Bull.* **37**:2498–2512.

Shearman, D. J., 1966, Origin of marine evaporites by diagenesis, *Inst. Mining and Metallurgy Trans.* **75**, sect. **B**:208–215.

Sloss, L. L., 1953, The significance of evaporites, *Jour. Sed. Petrology* **23**:143–161.

Stewart, F. H., 1963, Marine evaporites, U.S. Geol. Survey Prof. Paper 440-y, 53p.

Iron in Sedimentation

Iron minerals are one of the major indicators of chemical conditions in sediments. They are also one of the main coloring agents. If detrital, they may be unmodified indicators of provenance; however, they are commonly oxidized or reduced in the environment of weathering, during transport, or in the depositional setting, in which case the assemblage of iron-bearing minerals reflect paleoclimatic factors as these affect environmental chemistry. Many also are formed in the depositional (e.g., glauconite) or diagenetic (e.g., pyrite) environment, and indicate significant characteristics of these environments. Presence of organic matter (see Bass Becking and Moore, 1950), sulfur (e.g., originally as organic sulfates; see Curtis and Spears, 1968), and bacterial activity are particularly influential with respect to the origin and/or alteration of iron minerals.

Iron-rich sediments are those with more than 15% iron. *Ironstones* are iron-rich sediments of generally post-Precambrian age, up to tens of meters thick and often composed of replaced oölites or fossil debris (e.g., see Hunter, 1970). *Iron formations* are mostly Precambrian iron-rich sediments interlayered with silica-rich sediments (cherts). These are commonly hundreds of meters thick (to 600+ m), and usually have abundant iron silicates (see Trendall, 1968; James and Sims, 1973; Chauvel and Dimroth, 1974). Many unresolved problems remain in accounting for their origin. *Blacksands, bog iron ores,* and *gossans* are other iron-rich sediments. Table 29 lists the iron-rich minerals that are found in sedimentary deposits (see also James, 1966).

Iron occurs in two valence states: Fe^{2+} (the ferrous ion), which is soluble, and Fe^{3+} (the ferric ion), which is insoluble. Where oxygen is available, the oxidized form will be produced (and bound to the oxygen) unless the iron is held within crystallographic structures. As iron-bearing crystals weather, the iron released may be oxidized, and it is thus common to find such minerals with oxide coatings. Iron is transported in solution (Fe^{2+} variety), in organic matter (hemoglobin and chlorophyll), within detrital (e.g., ferromagnesian) minerals, as insoluble iron oxides coating grains, or adsorbed on clay minerals (see Carroll, 1958). It is prone to remobilization under reducing depositional or diagenetic conditions. Stability fields for the iron ions and oxides are shown in Fig 98 (page **190**); note that both Eh and pH are important. The influence of sulfide ions in the depositional environment is shown in Fig. 99 (page **191**). Availability of CO_2 is significant in the production of iron carbonates (together with an absence of sulfide ions). Iron silicate minerals are formed under conditions not fully understood, although quite a bit is known about glauconite (see section on "Clay Minerals"). Fig. 100 (page **192**) illustrates the diagenetic sequence of modifications that is inferred for siliceous iron formations.

The origin and significance of red beds in the geological column has been subject to much debate (see Krynine, 1948; Walker, 1967, 1974; Van Houten, 1973; Folk 1976; Ziegler and McKerrow, 1975). A multiplicity of origins must be accepted, but there may be evidence in the rock or associated facies for the explanation in any particular case.

Table 29. Iron-bearing Minerals

Oxides

Hematite, αFe_2O_3

Hexagonal. Principal mineral in Precambrian iron formations; also common in Lower Paleozoic ores; only locally important in Mesozoic and Tertiary ironstones, except as a product of secondary enrichment. May comprise almost 100% of rock. Main red coloring agent in rocks, but need not be abundant to produce distinctive, bright color. Called *martite* when occurring as octahedral pseudomorphs after magnetite.

Goethite, $\alpha FeO(OH)$ or ($\alpha Fe_2O_3 \cdot H_2O$?)

Orthorhombic. Brown, submetallic, crystalline but often botryoidal. Common in soils (brown color). Absent in Precambrian rocks, but may have been present and have converted to hematite. Common associate of chamosite, to which or from which it may convert.

Lepidocrocite, $\gamma FeO(OH)$ or $\gamma Fe_2O_3 \cdot H_2O$

Orthorhombic. Hydrated polymorph of goethite; uncommon. Red, brown, or yellow.

Ilmenite, $FeTiO_3$

Hexagonal-rhombohedral. If pure, contains 56% TiO_2; generally has less because of substitutions for Ti. Uncertain whether it forms under sedimentary conditions. May dominate blacksands.

Magnetite, $FeO \cdot Fe_2O_3$

Isometric. In Phanerozoic rocks, mostly of metamorphic origin but some is diagenetic or supergene. In Precambrian, abundant in association with Fe-silicate facies, apparently of early diagenetic origin. Formation requires intermediate Eh and high pH, possibly forms from dehydration of $Fe(OH)_2$ — see Fig. 98. Commonly in solid solution with ulvospinel (e.g., titanomagnetites of New Zealand).

Maghemite, γFe_2O_3

Isometric. Magnetic dimorph of hematite. Thought to be rare in nature but is abundant in some laterites. Possibly derived from oxidation of magnetite. Crystallography and x-ray pattern similar to magnetite. Generally black-blue gray.

Limonite, a mixture of iron oxides (esp. goethite and lepidocrocite and *turgite* — $Fe_2O_3 \cdot H_2O$)

Characteristic of bog iron ores and gossans, and locally mined. Generally earthy, yellow to dark brown. Can be formed and mobilized almost any time in the history of a rock.

Carbonates

Siderite, $FeCO_3$

Hexagonal-rhombohedral. The only common Fe-carbonate, although *ankerite* (Fe-rich dolomite) is locally abundant. Easily oxidized, whereupon commonly converts to mixture of limonite and calcite (common weathering or groundwater phenomenon). See Fig. 99 for optimal genetic conditions and Sellwood (1971) for an example of an occurrence. Important constituent of ironstones, where it generally is very finely crystalline and intimately intermixed with other minerals. Rarely economic because of impurities. Common accessory mineral in coal measures (e.g., as concretions, but also as "black bands" with carbonaceous or bituminous admixtures).

188

Table 29. *(Continued)*

Silicates

Glauconite, $(OH)_2K_{(x+y)}(Fe^{+3}, Al, Fe^{+2}, Mg)_{\sim 2}(Si_{(4-x)}Al_x)_{\sim 4}O_{10}$
where x = 0.2–0.6 and y = 0.4–0.6

Clay mineral. Not economic to date, but may become a K-fertilizer if cation-exchange capability can be enhanced. Very uncommon in Precambrian (only present in upper Proterozoic). *Not* a major component of iron formations or ironstones, although can comprise up to 75% of beds to more than 10m thick. Variety of parent materials; formed *almost* exclusively in marine environments (see expanded discussion in section on "Clay Minerals").

Chamosite, $?(OH)_8(Fe_4^{+2}AL_2)(Si_2Al_2)O_{10}$

Septechlorite. Commonest Fe-silicate in post-Precambrian iron formations and ironstones. Commonly oölitic, otherwise similar occurrence and field properties to glauconite. Primary mineral, but can alter to or from goethite and hematite and chlorite. Economic ore in some places (e.g., Jurassic of Britain and Alsace-Lorraine). Commonly associated with siderite and/or calcite. Marine or nonmarine (e.g., in the lower parts of peat horizons or lagoons). Forming today, probably under conditions subtly different from glauconite (for example, see Rohrlich et. al., 1969).

Greenalite, $Fe_3^{+2}(OH)_4Si_2O_5 \cdot H_2O$

Septechlorite analogous to antigorite. Dominantly in Precambrian (abundant); very rare in Paleozoic and Mesozoic. Typically associated with magnetite. Resembles chamosite and glauconite but is isotropic when transparent in thin section.

Thuringite, $(OH)_{16}(Mg_{1.4}Fe_{7.4}^{+2}Fe_{1.5}^{+3}Al_{1.7})(Si_{4.8}Al_{3.2})O_{20}$

Fe-rich chlorite. Locally abundant in Paleozoic ironstones; may be oölitic. Diagenetic or metamorphic alteration of other Fe-silicates.

Minnesotaite, $(OH)_{11}(Fe^{+2}Mg)_{11}(SiAlFe^{+2})_{16}O_{37}$

Fe-analogue of talc. Greenish gray, waxy. May be primary in Precambrian (?), mostly is metamorphic in iron formations.

Stilpnomelane, $(OH)_4(KNaCa)_{0-1}(Fe,Mg,Al)_{7-8}Si_8O_{23-24}(H_2O)_{2-4}$

"Brittle mica." Very similar to biotite (and probably often misidentified). May be primary, often *is* metamorphic. Precambrian *and* younger.

Sulfides

Pyrite, FeS_2

Isometric. Mainly an accessory, diagenetic mineral in sediments, disseminated or concentrated in nodular crystalline aggregates or replacing fossils. Locally to 65%+ of layers to 0.5m thick in black shales, where it may be associated with other economic metal sulfides (for example, see Love and Zimmerman, 1961). Microspheres about or associated with bacterial sulfate-reducers in modern and ancient muds. Formed under reducing conditions in association generally with organic material (see Fig. 99).

Marcasite, FeS_2

Orthorhombic. Dimorph of pyrite. Commonly associated with coals. Apparently formed under reducing and acidic conditions. Less stable than pyrite; oxidizes easily. Commonly associated with siderite. Sometimes after pyrrhotite in lowest weathering zones of ore deposits.

(Continues on next page)

Table 29. *(Continued)*

Pyrrhotite, $Fe_{1-x}S$
  Rare in sediments, usually then as oölites (of replacement origin?) with phosphatic nodules or siderite. In some recent muds.

Hydrotroilite, amorphous $FeS \cdot nH_2O$? Highly metastable
  and
Melnikovite, amorphous mixture with Fe_3S_4 a major component.
  Both are black colloids, the latter common in Recent and Tertiary rocks and sediments. Color some sediments and may form small nodules (1 mm). Very difficult to identify, but may be common. Apparently convert readily to pyrite or other sufides. Melnikovite sometimes gives pyrite, sometimes marcasite x-ray pattern.

Phosphates

Vivianite, $Fe_3P_2O_8 \cdot 8H_2O$
  Bright blue colors distinctive. Occurs in bogs, swamps, and after wood fragments.

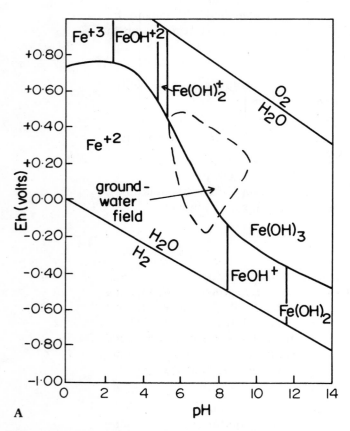

A

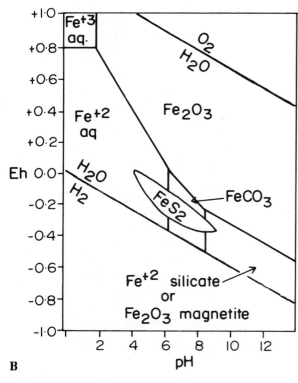

B

Figure 98.   Simple Eh/pH phase diagrams for sedimentary iron minerals. *A:* Stability fields for aqueous ferrous-ferric system in terms of Eh and pH. *(After Walker, 1967)*   *B:* Stability fields for iron minerals in water at 25°C, 1 atmosphere, $CO_2 = 10^0$, sulfur 10^{-6}m; quantity of dissolved silica present probably determines whether magnetite or ferrous silicate forms. *(After Garrels and Christ, 1965)*

Stability fields in marine depositional waters at pH8, solid
stability at $a_{Fe+2} < 10^{-6}$, as a function of Eh and
Upper: HS⁻ activity, with $a_{HCO_3^-} = 10^{-7}$
Lower: HCO₃⁻ activity, with $a_{HS^-} = 10^{-7}$

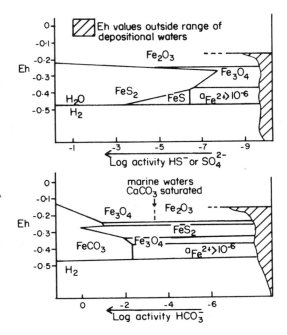

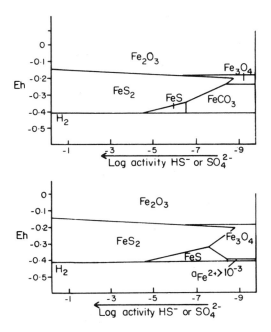

Mineral stability fields as a function of Eh and HS⁻ activity,
pH7, solid stability at $a_{Fe+2} < 10^{-3}$
Upper: probable conditions of restricted pore water
circulation, $a_{HCO_3^-} = 10^{-2.5}$
Lower: conditions of low carbonate activity, $a_{HCO_3^-} = 10^{-4.4}$

Stability fields as a function of Eh and HS⁻ activity; solid
stability at $a_{Fe+2} < 10^{-6}$
Upper: anoxic marine waters, pH8, with SO₄⁻²/HS⁻ equilibrium
disturbed in favor of HS⁻ and $a_{HCO_3^-} = 10^{-3.5}$
Lower: probable upper sediment pore waters, pH7, $a_{HCO_3^-} = 10^{-2.5}$

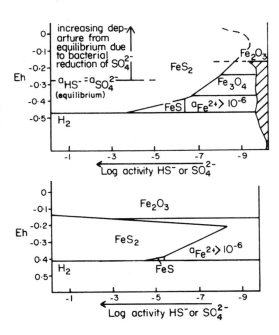

Figure 99. Influence of sulfide and bicarbonate
ions on iron mineralogy in sedimentary environments. *Major conclusions* (Curtis and Spears):
Ferric compounds (e.g., hematite) are the only iron
minerals that can exist in true equilibrium with
depositional waters. Ferrous minerals can only attain
equilibrium with sediment pore waters (i.e., are only
stable within sediment masses), except for *pyrite,*
which is stable relative to all other possible phases
even in the presence of low-sulfide activities. Pyrite
is a metastable phase in anoxic water masses where
bacteria maintain sulfide activities at nonequilibrium
levels. Optimum conditions for siderite are severely
restricted circulation, zero sulfide activity, and low
Eh (−0.25 to −0.35v.); ferrous silicates are stable
in similar conditions where bicarbonate activity is
low and there is saturation with some active silica
form. *(After Curtis and Spears, 1968)*

191

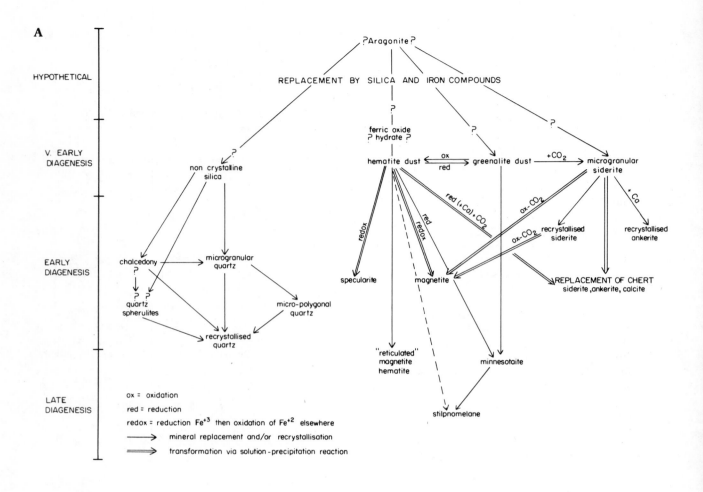

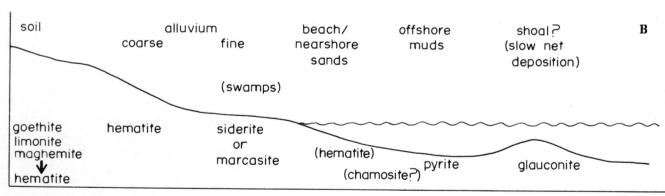

Figure 100. Diagenetic facies of iron formation. A: Succession of changes suggested for Precambrian iron formations of the Labrador Trough by Dimroth. *(After Dimroth in Walker, 1979)* B: Simplistic sketch illustrating relationships between depositional environment and characteristic diagenetic iron compounds.

LITERATURE

Bass Becking, L. G. M., and D. Moore, 1950, The relation between iron and organic matter in sediments, *Jour. Sed. Petrology* **29**:454-458.

Carroll, D., 1958, Role of clay minerals in the transportation of iron, *Geochim. et Cosmochim. Acta* **14**:1-27.

Chauvel, J-J., and E. Dimroth, 1974, Facies types and depositional environment of the Sokoman Iron Formation, Central Labrador Trough, Quebec, Canada, *Jour. Sed. Petrology* **44**:299-327.

Curtis, C. D., and D. A. Spears, 1968, The formation of sedimentary iron minerals, *Econ. Geology* **63**:257-270.

Dimroth, E., 1979, Diagenetic facies of iron formation: in R. G. Walker (ed.), *Facies Models*, Geological Association of Canada Reprint series no. 1, Geoscience Canada, Toronto, 183-190.

Drever, J. I., 1974, Geochemical model for the origin of Precambrian banded iron formations, *Geol. Soc. America Bull.* **85**:1099-1106.

Folk, R. L., 1976, Reddening of desert sands: Simpson Desert, N.T., Australia, *Jour Sed. Petrology* **46**:604-615.

Hunter, R. E., 1970, Facies of iron sedimentation in the Clinton Group, in G. W. Fisher, F. J. Pettijohn, J. C. Reed, Jr., and K. N. Weaver (eds.), *Studies of Appalachian Geology, Central and Southern*, Wiley-Interscience, New York, 101-121.

James, H. L., 1966, Chemistry of the iron-rich sedimentary rocks. *U. S. Geol. Survey Prof. Paper 440W*, 61p.

James, H. L., and P. K. Sims (eds.), 1973, Precambrian iron-formations of the world, *Econ. Geol.* **68**:913-1221.

Krynine, P. D., 1948, The origin of red beds, *New York Acad. Sci. Trans.* **11**:60-68.

Lepp, H. (ed.), 1975, *Geochemistry of Iron*, Benchmark Papers in Geology, vol. 18, Dowden, Hutchinson & Ross, Stroudsburg, Pa., 464p.

Lepp, H., and S. S. Goldich, 1964, Origin of Precambrian iron formations, *Econ. Geology* **59**:1025-1060.

Love, L. G., and D. O. Zimmerman, 1961, Bedded pyrite and microorganisms from Mount Isa shale, *Econ. Geology* **56**:873-896.

Rohrlich, V., N. B. Price, and S. E. Calvert, 1969, Chamosite in the recent sediments of Loch Etive, Scotland, *Jour. Sed. Petrology* **39**:624-631.

Sellwood, B. W., 1971, The genesis of some sideritic beds in the Yorkshire Lias (England), *Jour. Sed. Petrology* **41**:854-858.

Trendall, A. F., 1968, Three great basins of Pre-Cambrian iron formation deposition: a systematic comparison, *Geol. Soc. America Bull.* **79**:1527-1544.

van Houten, F. B., 1963, Origin of red beds: a review— 1961-1972, *Ann. Rev. Earth and Planet. Sci.* **1**:39-61.

Walker, T. R., 1967, Formation of red beds in modern and ancient deserts, *Geol. Soc. America Bull.* **78**:353-368. (Also see discussion by R. F. Schmalz, 1968, *Geol. Soc. America Bull.* **79**:277-280, and reply by Walker, 281-282.)

Walker, T. R., 1974, Formation of red beds in moist tropical climates: a hypothesis, *Geol. Soc. America Bull.* **85**:633-638.

Ziegler, A. M., and W. S. McKerrow, 1975, Silurian marine red beds, *Am. Jour. Sci.* **275**:31-57.

Introduction to Well Logging

Petroleum geologists spend much of their time working with geophysical data such as gravity and magnetic maps and seismograms, but all spend at least some of their time at the drilling rig (Fig. 101). Well logs are the principal tools for studying subsurface rocks and their interstitial fluids. During the drilling operation, the driller prepares logs of drilling rate while the well-site geologist logs lithology from well cuttings (small chips cut by the bit and brought to the surface in the drilling mud, from which they are separated on a vibrating screen). Gas detector and gas chromatograph readings are taken, and sometimes other data (such as sample fluorescence). Micropaleontologists examine selected samples to provide stratigraphic control, which is also shown on the lithological logs. Well cores may be taken over specific intervals —but rarely, because the cost is high. After drilling, contracted specialists lower complex instrument packages into the well and compile wireline logs.

The wireline logs provide information about the *in situ* characteristics of the formations penetrated by the well. Various physical parameters are measured, such as *bulk properties of the rock*—shown on dipmeter, sonic, and formation-density-compensated (FDC) logs; *compositional aspects of the rock* —shown on gamma-ray logs; and *fluid properties in the rock pores*—shown on self-potential (SP), resistivity (R), and compensated neutron logs (CNL). In addition, some instruments record characteristics of the well hole itself and its casing (if any)— results are of primary interest to engineers and only the caliper log will be discussed below. If warranted, additional tests (such as for fluid recovery and formation pressure) may be made or further samples collected (such as sidewall cores).

The total of all the log information is integrated with data from gravity, magnetic, and seismic surveys, regional geological knowledge, other nearby wells, analogous sequences elsewhere, detailed study of rock samples, and, if relevant, analysis of petroleum transport and market conditions, to produce a final report by the geologist to his or her employer.

INSTRUMENTS AND WELL LOG VARIETIES

The *caliper* yields information that is not directly relevant to either geology or hydrocarbon potential; it measures only the borehole diameter. This information is important, however, because the condition of the borehole affects the response of most logging tools. Thus, a caliper log is used to check the reliability of other logging measurements.

Borehole diameter reflects the ability of the rock to stand unsupported. Tough homogeneous rock (such as some limestones) will have a hole diameter close to the size of the drilling bit; shales may have holes twice the bit diameter because of caving.

Resistivity logs depict the resistance of a formation to an applied electrical current. The current is held constant and resistivity is calculated from measurements of changes in voltage via Ohm's law:

$$\text{Voltage} = I \text{ (current)} \times \text{Resistivity}.$$

The main purpose of resistivity measurements is to detect any abnormally high resistance imparted to formations by the presence of interstitial hydrocarbons. Because drilling mud frequently invades

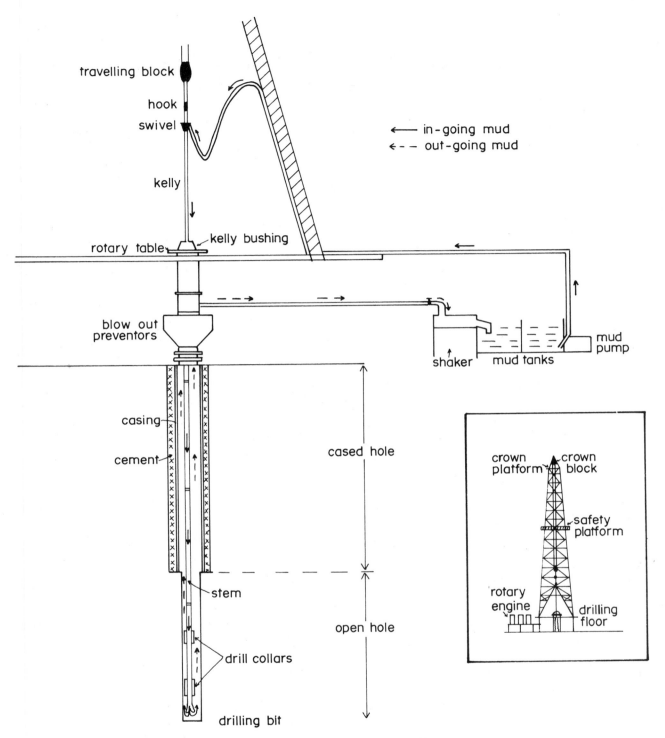

Figure 101. Major features of a drilling rig.

195

permeable rocks, and thereby affects their resistivity, the instrument measures and contrasts the resistivity of the invaded (flushed) zone with that of the uninvaded zone (see Fig. 102). Because normally R_m (mud resistivity) is greater than R_w (resistivity of formation fluid), the resistivity of the flushed zone (R_{xo}) is generally greater than that of the uninvaded zone (R_o). Exceptions occur where oil and gas are present, when R_o will equal or exceed R_{xo}.

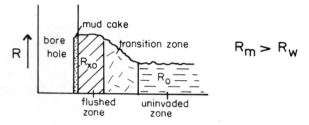

Figure 102.   Contrasting resistivity measurements.

Electrodes are lowered down the well and the voltage difference between them is measured when an outside current is sent into the wall rock. The depth of penetration of the measurement into the rock depends on the electrode spacing. Close electrode spacing measures R_{xo} and provides the short normal, or *laterolog*, curve. R_o is measured by a deep investigation tool—most commonly these days, an induction device that consists of a series of coils that generate, impart, and detect magnetic fields from alternating electric currents. The depth to which the device measures in the formation is about 325 cm. The results are plotted as the *induction* curve. The induction device is used in preference to widely spread electrodes (which give "long normal" curves) because the electric energy can be beamed directly into the rocks horizontally, thus it achieves greater penetration and measures resistivity of a thinner layer. In addition the induction measure is independent of the conductivity of the drilling fluid (if nonconductive, oil-based muds are used, the electrode instrument cannot be used). A rock with 10% porosity is 10 times as resistive as one with 30% porosity if the same pore fluid is present.

Resistivity values may be used to calculate S_w, a theoretical value representing percentage pore space occupied by water; this value is important in evaluating hydrocarbon content.

A *microlog* is sometimes run to measure the thickness of the mud cake along the borehole. Thickness of mud caking reflects the permeability (and porosity) of the strata—impermeable horizons have

a thin (e.g., 0.8 mm) caking, whereas permeable horizons have a relatively thick (e.g., 12.7 mm) caking. Caliper logs are taken in conjunction with the micrologs. The microlog tool is a resistivity device with very small spacing between electrodes.

The *dipmeter* tool is a high-resolution resistivity device with four electrodes at 90°. It provides four continuous logs that are used to correlate thin intervals of equal resistivity across the borehole. The assumption is made that intervals of equal resistivity are the same lithological unit. Computer selection of similar resistivities on the four curves allows calculation of direction and amount of dip in relation to magnetic north (also recorded during logging). The attitude of internal stratification is shown, as well as the attitude of the strata as a whole; hence these logs can be used for paleoenvironmental studies based on sedimentary structures, as well as for determining tectonic deformation and the presence of angular unconformities. See Fig. 103.

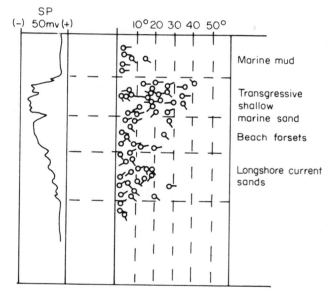

Figure 103.   Environmental interpretation using the high-resolution dipmeter. *(After Gilreath et al., 1969)*

Spontaneous potential (SP) logs are used to interpret electrical potential differences, which reflect the permeability of rocks. Measurement is made of voltage differences between a reference electrode at the surface and a movable electrode in the well; generally these differences are recorded in conjunction with deep resistivity measurements. Voltage differences result mainly (85%) from the movement of ions from the formation into the borehole mud (*electrochemical effect*) and partly (15%) from the invasion of the drilling mud into the formation (*electrofiltration effect*). The electrochemical effect is generated primarily by the *shale membrane potential*. Clay minerals within shales permit passage of Na^+ ions, but not Cl^- ions, from more saline fluids to less saline fluids. In the case of a borehole filled with drilling mud, the Na^+ ions will move from the saline waters of a porous bed (such as a sandstone) through the shale to the less saline drilling mud (see Fig. 104). There is also a less significant movement of Cl^- ions directly to the drilling mud from the formation fluids where they are in contact (*liquid junction potential*). A natural electric current results from these effects, and the resultant voltages are measured and logged. The electrofiltration effect is generated by the flow of electrolyte filtrate from the drilling fluid through the mud cake into a permeable formation (an electrokinetic force is generated by the fluid movement that results from the differential pressure between fluids in the well and the formation). Only in unusual circumstances is this effect significant in SP log deflections.

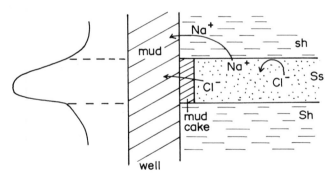

Figure 104. The shale membrane potential. *Note:* Opposite shales, the shape of the SP curve tends to be constant; the straight line on the log is the "shale base line."

For SP curves to be generated, the drilling must be conductive, the salinity of the drilling mud must be less than that of the formation fluids, and the drilling mud must invade the formation. If the resistivity of the mud (R_m) equals the resistivity of the formation fluid (R_w), there is no response (Fig. 107, level H); if R_m is less than R_w (as in the case of a freshwater sand), there will be a negative SP response (Fig. 107, level B). When the rocks are impervious, such as "tight" sandstones (Fig. 107, level L), evaporites, and many carbonates (level J), there is no SP response. Other data must be used in such

cases to determine the lithology. Because SP responds to the salinity of the formation fluids, it can be used to define their resistivity qualitatively (see later).

Sonic logs measure transit times of P-waves through rock intervals. High-density rocks promote faster P-wave velocities and therefore shorter transit times than low-density rocks (see Fig. 105). Because density can be related to lithology, sonic logs are used for lithological differentiation in conjunction with other well logs. They are particularly useful in identifying specific horizons which, because of a marked density contrast with the associated strata, have acted as reflectors in seismic reflection profiles. Where the lithology is known, and because porosity influences density, the sonic log can be used to interpret porosity; for example, in Fig. 107, the decrease in porosity with depth is reflected by shorter transit time (that is, lesser deflection—compare interval E-M with A-E).

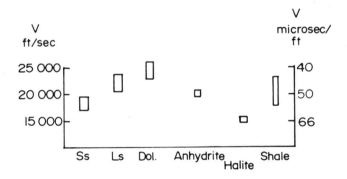

Figure 105. Example of some acoustic velocity ranges for sediments.

Gamma-ray logs measure the natural radioactivity of the strata, and are used for lithological identification and correlation. Because gamma rays have high energy, these logs can be run where the well is cased (many other geophysical measurements can be made only in open holes). The radiation is produced by U-238, U-235, Th-232, and K-40, and since the first three isotopes are rare in nature, the K-40 source principally dictates the character of the logs. Potassium is abundant in alkali feldspars, micas, clays, and organic compounds (not hydrocarbons in general). Hence, where organic matter is negligible, quartz sandstones and carbonates should produce the least radiation and shales the most. However, where organic matter is present or the sandstones have abundant K-feldspars, mica, and/or clayey matrix, relatively high gamma values may be recorded (see Fig. 106). Other well logs must be used in these cases to distinguish lithology (such as the resistivity log in Fig. 107, level D). In coal-bearing sequences, these logs are particularly useful because of the high gamma radiation emitted by concentrations of organic matter. In sandstone/shale sequences, the logs generally look similar to SP logs.

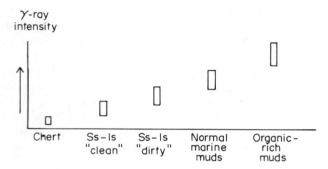

Figure 106.   Expected relative ranges of gamma-ray intensities for some common sediments.

Formation-density compensated logs (FDCs) measure the electron density of formations by bombarding them with gamma rays and analyzing the resultant scattered radiation. The electron density is proportional to the bulk density, thus the curve depicts density changes. Insofar as porosity has a major influence on density, these logs can be used to calculate porosity ($\emptyset$) by:

$$\emptyset = \frac{\rho_{ma} - \rho_b}{\rho_{ma} - \rho_f}$$

where ρ_b = bulk density (generally taken as the apparent density as read from the log); ρ_{ma} = matrix density (measured separately); and ρ_f = density of mobile fluid in the rock pores (measured separately).

So calculated, $\emptyset$ is the *effective porosity*; unlike the case for porosity derived from sonic logs, the water held tightly by clay minerals in the rock is generally recorded as part of the matrix rather than as open pore space. Excessively large porosities will result from the calculations if appreciable gas is present.

Some average density values are: shales, 2.2–2.75; sandstone, 2.4–2.6; carbonates, 2.5–2.9, halite, 2.0; anhydrite, 3.0; coal, 1–2.0.

Compensated neutron logs (CNLs) measure the gamma radiation emitted when a formation is bombarded by neutrons. High-energy neutrons are emitted by the instrument source and lose energy as they collide with nuclei in the formation; they may then be captured by the nuclei, which become highly excited in the process and emit gamma rays. Neutrons lose most energy when they collide with hydrogen; hence the intensity of gamma radiation recorded is a measure of hydrogen concentration. Hydrogen is concentrated in formation fluids, and the quantity of fluids present is a function of porosity; the neutron logs, therefore, can be used to determine porosity (but not permeability). The apparent porosity is reduced by the presence of appreciable gas, but not much by the presence of oil. CNL and FDC curves are commonly plotted on the same log because the presence of hydrocarbons is generally indicated where the two curves approach or overlap (see Fig. 107, level F). As with gamma-ray logs, neutron logs can be made in cased wells. Fluids contained in the borehole will muffle the effect of neutron bombardment of the wall rock, thus caliper logs should be studied along with the CNLs.

GENERAL PROBLEMS

The chemistry of the drilling fluid is critical. Water-based muds conduct electricity, whereas oil- or air-based ones do not; the salinity of the mud-water system influences conductivity. A fluid with known conductivity is necessary for the production and interpretation of SP and R logs. Mixing of water from the mud with fresh formation waters, particularly at great depths, may result in a lowering of initial drilling mud salinity, or mixing with more saline formation waters may increase initial salinity of the muds and reduce salinity in the formation with a consequent swelling of the formation clays— the geophysical response of the formation changes, permeability may be destroyed (ruining production potential), and the clays may even swell sufficiently to trap the drilling pipe and force abandonment of the well. Where seawater (or oil) is used to prepare the drilling mud, as on many offshore rigs, SP measurements cannot be made.

Irregular configuration of the borehole, excessive invasion of the formation by the drilling mud (for example, due to excessive weight in the mud column), and the thickness of the mud cake can all adversely affect some geophysical log data.

Rising temperatures decrease resistivity values and thus influence the interpretation of R logs.

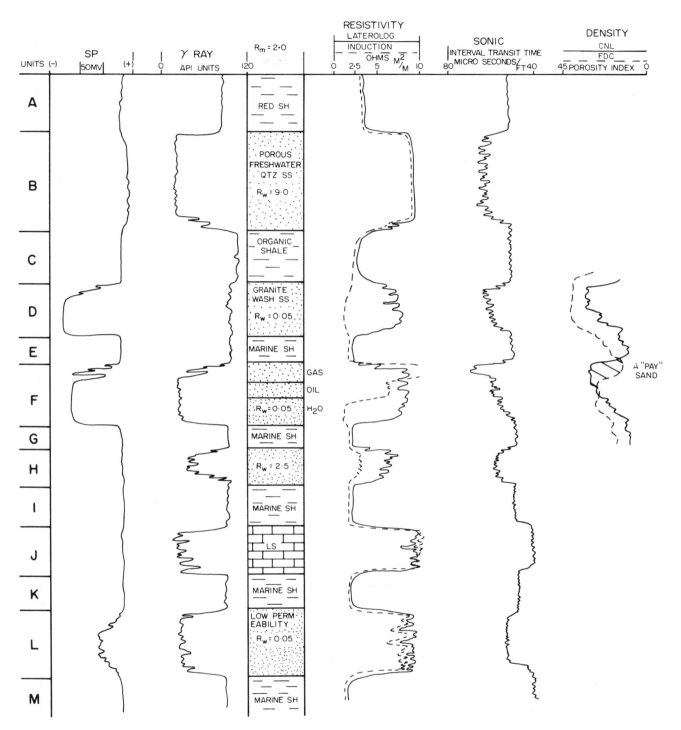

Figure 107. Idealized well log responses.

199

SEISMIC INVESTIGATIONS

Although it is beyond the scope of this manual to describe or discuss principles, procedures, and interpretations, it is important that the sedimentary geologist be aware of the extensive contributions that are made in subsurface exploration by analysis of seismic data. Preliminary interpretations of subsurface structure and stratigraphy are generally made before wells are drilled. Well logs provide control data for expanding and refining the interpretation of seismic records. The combination of data and interpolation lead to the construction of subsurface maps (see separate chapter), which depict characteristics of the stratigraphic succession over a wide area.

A modern review of the applications of seismic investigations is provided in Payton (1977).

INTERPRETATIONS FROM GEOPHYSICAL DATA

Lithology

Within one field the same strata tend to have similar geophysical properties, but significant variation can result from the presence of different interstitial fluids, different degrees of compaction/cementation, and even different time intervals between the drilling and the logging procedures. Not only may rocks that appear petrographically identical log very differently, but rocks that appear different may respond very similarly to a particular geophysical measurement. Hence a number of different well logs must be considered jointly for lithological interpretation, and must be integrated with any other data available (the well cuttings log, fossils, seismic profiles, prior knowledge of the basin, and so on).

Depositional environment is inferred from the rock type, the three-dimensional geometry of the unit, the vertical succession (in comparison with ideal vertical profile models determined from previous work on stratigraphic successions), lateral facies relationships, and paleontological data. Fig. 108 shows characteristics of selected "ideal" stratigraphic sequences. Boundaries between stratigraphic units are found primarily from the SP or gamma-ray logs; depths so determined are used in constructing isopach and structural contour maps that depict the three-dimensional geometry. Data from well cuttings, SP logs, and gamma-ray logs, together with R logs, help define the lithologies involved. Dipmeter readings can be useful in detecting some characteristic forms of internal stratification (such as cross-bedding; a large scatter of dip magnitudes is common in high-energy deposits such as bars or dunes), and may be useful in determining paleocurrent directions. In addition, SP logs can be used to interpret texture—the coarser the grain size, the greater the "kick" in otherwise similar units—see Fig. 108, (c) and (d). Note that there is an inherent limitation to single well log interpretations—the profile represents the vertical succession along a single vertical line. The three-dimensional geometry of a unit is *inferred* from correlation between wells; miscorrelation is easy, especially if a number of similar strata occur in a sequence (for example, correlation between two different fluvial channels could produce a false impression of a single-sheet sandstone unit). Units characteristic of "ideal" vertical profile models may also be absent and lead to misinterpretation, for nature quite commonly deviates from the ideal!

Resistivity logs, together with SP logs, can depict trends of porosity and permeability (if cores are available for laboratory testing, much greater use can be made of the data). The trends, if within single lithological units, can be used to interpret diagenetic effects (such as increasing compaction or cementation). Where resistivity is high in both deep and shallow logs (see Fig. 107, level J), low-porosity ("tight") rocks may be indicated; if SP development is low, low porosity is confirmed (since the drilling mud must invade the formation for a response). Neither of these logs, however, may distinguish well-cemented sandstones from impervious carbonates.

Hydrocarbons

Quantitative analysis of geophysical log data provides values of porosity and degree of hydrocarbon saturation. Porosity is primarily found from sonic and density logs, but CNL also can be calibrated to give porosity values. For hydrocarbon evaluation, the percentage of pore space occupied by water in the uninvaded zone (S_w) is compared with the residual water saturation in the invaded

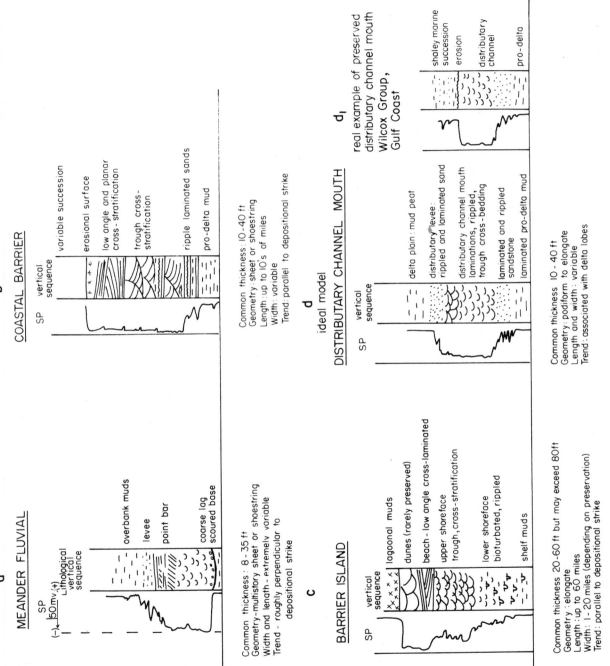

Figure 108. Idealized examples of stratigraphic motifs and self-potential logs. (*After Galloway, 1978*)

a

MEANDER FLUVIAL

SP   Lithological
(-) [50 mv](+) vertical
              sequence

overbank muds
levee
point bar
coarse lag
scoured base

Common thickness: 8 - 35 ft
Geometry-multistory sheet or shoestring
Width and length - extremely variable
Trend - roughly perpendicular to
    depositional strike

b

COASTAL BARRIER

SP   vertical
      sequence

variable succession
erosional surface
low angle and planar
cross - stratification
trough cross-
    stratification
ripple laminated sands
pro-delta mud

Common thickness: 10-40 ft
Geometry: sheet or shoestring
Length: up to 10's of miles
Width: variable
Trend: parallel to depositional strike

c

BARRIER ISLAND

SP   vertical
      sequence

lagoonal muds
dunes (rarely preserved)
beach - low angle cross-laminated
upper shoreface
trough, cross-stratification
lower shoreface
bioturbated, rippled
shelf muds

Common thickness 20-60 ft but may exceed 80ft
Geometry : elongate
Length : up to 60 miles
Width : 1- 20 miles (depending on preservation)
Trend : parallel to depositional strike

d

ideal model

DISTRIBUTARY CHANNEL MOUTH

SP   vertical
      sequence

delta plain : mud peat
distributary levee:
rippled and laminated sand
distributary channel mouth
laminations, rippled,
trough cross-bedding
laminated and rippled
sandstone
laminated pro-delta mud

Common thickness 10 - 40 ft
Geometry: podiform to elongate
Length and width: variable
Trend: associated with delta lobes

d₁

real example of preserved
distributary channel mouth
Wilcox Group,
Gulf Coast

shaley marine
succession
erosion
distributary
channel
pro-delta

zone (S_{xo}). Pore space occupied by hydrocarbons will be $1/S_w$, and that occupied by residual oil is $1/S_{xo}$. S_w is calculated by the Archie equation:

$$S_w{}^x = FR_w/R_t$$

where R_w = resistivity of formation fluid (found from the SP data); R_t = deep induction reading (R_o is assumed to approximate R_t); x generally approximates 2; and F (which is proportional to porosity) is about $1/2$ in carbonates and $0.81/2$ in sandstones, but should be measured separately in each situation. $S_{xo}{}^2 = FR_{mf}/R_{xo}$ where R_{mf} is the resistivity of the mud cake and R_{xo} is the resistivity of the invaded zone (obtained from laterolog).

The percentage of hydrocarbons that can potentially be produced is $S_{xo} - S_w$.

Coal

Wireline logs accurately define seam thickness and provide excellent data for regional correlation (conventional coring often does not recover the full interval and conventional well-cutting techniques are inaccurate). Evaluation of moisture content and ash content can be made from density and neutron logs after initial calibration in each region. The FDC and sonic (both P and S wave) logs can be used to determine Young's modulus and Poisson's ratio, and they in turn provide information about the strength of roof and floor rocks, which is useful when planning mining operations. The gamma-ray logs are particularly important in defining seam geometry.

INTERPRETATION OF PALEOENVIRONMENTS FROM SMALL SAMPLES

Stratigraphic associations and relationships, three-dimensional geometry of successions and specific lithological units, and sedimentary structures are major considerations in paleoenvironmental studies and require fieldwork, borehole, and/or geophysical data. Hand specimen and laboratory analyses should not be expected to provide definitive evidence of paleoenvironments. However, small samples of sediments must be examined when making or refining interpretations and may be all that you can obtain of the actual deposit. Some of the salient characteristics that may be useful are discussed here.

Color: Color may be imparted by the aggregate color of transported particles or by materials introduced after deposition by chemical or biochemical agencies. Red to yellow-brown colors generally reflect the presence of iron oxides, the distribution of which with respect to primary structures or grains may indicate whether they developed during diagenesis or were supplied with the sediments. (For example, Leisëgang bands cross primary structures and are clearly diagenetic; concentrations of iron oxide specks around detrital grains underneath clear overgrowths suggest the oxides were on the grains at the time of burial.) Only a small quantity of iron oxides (1-2%) may produce distinctive or bright colors. Black colors generally reflect the presence of organic matter and suggest that the depositional (and subsequent diagenetic) environment was not oxidizing. Some black coloration results from disseminated, very fine iron sulfides (similar environmental implications) or the presence of detrital magnetite/ilmenite (potential deposition in oxidizing environment). Mottling may reflect bioturbation or irregular, chemically caused oxidation and reduction.

Lamination: Bedform configurations are variously indicative of depositional/erosional processes, or of postdepositional modifications ranging from load deformation to laminar flow during quasi-fluid injection. Horizontal lamination may indicate intermittent deposition from passive suspension or from entrainment flows in the lower flow regime. Flat bedforms in upper flow regimes and gravity flows have been described but are unlikely in small samples. Cross-lamination reflects lower-flow-regime ripple bedforms; cross-bedding reflects more rapid currents, but still a lower flow regime (dune bedforms). See "Sedimentary Structures" for other bedforms.

Graded bedding: Normal graded bedding may reflect waning currents or "dumping" in a low-energy setting; the former should show moderate to good sorting in all horizontal planes, the latter should show improved sorting upward. Inverse grading (coarsening upwards) reflects a waxing of the current energy during deposition of the unit.

Burrows: Burrows indicate that bottom waters (not necessarily the sediments) were oxygenated, and that the substrate was not lithified. Distinctive varieties and relative abundances may be highly informative (see "Biogenic Structures").

Borings: When borings were made, the sediment was lithified and bottom waters were oxygenated. A disconformity is likely.

Fossils: Fossils are informative to the extent of existing knowledge concerning their paleoecology— but the extent of transport must be evaluated. Articulated shells are suggestive of *in situ* burial, but storm or gravity-flow events may redeposit living organisms without breakage (the character of the surrounding sediment should be indicative). A thick periostracum suggests a freshwater organism. The absence of fossils is not diagnostic.

Grain size: If the material was deposited from entrainment flow or from suspension (which you can evaluate from other data), see the Hjulstrom diagram (Fig. 44) for the *minimal* energy level of transportation and deposition. The degree of sorting generally reflects the relative activity of the depositing currents (or waves), but poor sorting may result from bioturbation, and the previous history of the sediment may restrict the size range available to the final depositional agencies. Quantitative analyses may be highly informative when data obtained are compared with the available data base, but generally you should not undertake it for process-interpretation from a single sample—data of value must be collected from a suite of samples from closely related strata. (See chapters in this manual on texture of detrital sediments and on sedimentary carbonates.)

Matrix content: If framework grains are matrix-supported (sandstones or conglomerates), they indicate rapid deposition in an environment with restricted winnowing. If grain-supported, matrix may be infiltrated after deposition, but the quantity is usually indicative of winnowing capability in the depositional setting. Matrix-supported sediments commonly reflect gravity flow deposition or bioturbation.

Roundness: Roundness of detrital particles reflects abrasion throughout the history of that grain —there may be no relation to the depositional environment. "Broken rounds" (originally rounded grains that are now angular due to fragmentation) are suggestive of high-energy conditions in the final environment. Soft minerals that are angular suggest little abrasion in the final environment. In nondetrital particles, the degree of abrasion may be indicated, but these should be evaluated in relation to their initial shape (some originate as rounded spherical particles) and internal structure (some shells will never become rounded).

Composition of clasts: See previous chapters in this manual.

Clay minerals: Cations adsorbed on the clays may be highly indicative of the chemistry of the depositional environment, but diagenetic chemistry must be considered also. In nonmarine deposits, clay mineralogy should be indicative of paleoclimatic/pedogenic conditions.

Glauconite pellets: These generally indicate a marine paleoenvironment, but insofar as they can be transported and even recycled final deposition in a nonmarine setting is possible (expect few and angular grains).

LITERATURE

Argall, G. O., Jnr., 1979, *Coal Exploration, 2nd International Coal Exploration Symposium, Denver, Proceedings,* vol. 2, Colorado, October 1978, Miller-Freeman, San Francisco, 560p.

Asquith, G., and C. Gibson, 1982, *Basic Well Log Analysis for Geologists, Methods in Exploration,* American Association of Petroleum Geologists/Society of Economic Paleontologists and Mineralogists, Tulsa, Okla., 216p.

Busch, D. A., 1974, *Stratigraphic traps in Sandstones— Exploration Techniques,* Am. Assoc. Petroleum Geologists Mem. 21, 174p.

Campbell, R. L., 1968, Stratigraphic applications of dipmeter data in the Mid-Continent, USA, *Am. Assoc. Petroleum Geologists Bull.* **52**:1700–1719.

Fons, L., 1969, Geological applications of well logs, *Society of Professional Well Log Analysts 10th Logging Symposium,* 1–41.

Galloway, W.E., 1978, *Exploration for Stratigraphic Traps in Terrigenous Clastic Depositional Systems,* Am. Assoc. Petroleum Geologists Short Course No. 3, lecture notes, Houston Geological Society.

Gilreath, J. A., J. S. Healy, and J. N. Yelverton, 1969, Depositional environments defined by dipmeter interpretation, *Gulf Coast Assoc. Geol. Socs. Trans.* **19**: 101–109.

Hobson, G. D., and W. Pohl (eds.), 1973, *Modern Petroleum Technology,* Applied Science Publishers, Barking, Essex, 996p.

Jaegler, A. H., and D. R. Matuszak, 1972, Use of well logs and dipmeter in stratigraphic-trap exploration, in R. E. King (ed.), *Stratigraphic Oil and Gas Fields,* Am. Assoc. Petroleum Geologists Mem. 16, 107–135.

Le Roy, L. W., D. O. Le Roy, and J. W. Raese, (eds.), 1977, *Subsurface Geology: Petroleum Mining Construction,* 4th ed., Colorado School of Mines, Golden, Colo., 941p.

Lynch, E. J., 1962, *Formation Evaluation,* Harper & Row, New York, 422p.

Merkel, R. H., 1981, *Well Log Formation Evaluation,* AAPG Continuing Education Series No. 14, American Association of Petroleum Geologists, Tulsa, Okla., 82p.

Moore, C. A., 1963, *Handbook of Subsurface Geology*, Harper & Row, New York, 235p.

Neidell, N. S., 1980, *Stratigraphic Modeling and Interpretation: Geophysical Principles and Techniques*, AAPG Continuing Education Series No. 13, American Association of Petroleum Geologists, Tulsa, Okla., 145p.

Payton, C. E. (ed.), 1977, *Seismic Stratigraphy—Applications to Hydrocarbon Exploration*, Am. Assoc. Petroleum Geologists Mem. 26, 516p.

Pirson, S. J. 1977, *Geologic Well Log Analysis*, 2nd ed., Gulf Publishing, Houston, Tex., 377p.

Ranson, R. C. (ed.), 1975, *Glossary of Terms and Expressions Used in Well Logging*, Society of Professional Well Log Analysts, 74p.

Schumberger Ltd., 1972, *Log Interpretation Principles*, 2 vols., Schlumberger, New York, 112p., 92p.

Selley, R. C., 1976, Subsurface environmental analysis of North Sea sediments, *Am. Assoc. Petroleum Geologists Bull.* **60**:184–195.

Selley, R. C., 1978, *Concepts and Methods of Subsurface Facies Analysis*, AAPG Continuing Education Series No. 9, American Association of Petroleum Geologists, Tulsa, Okla., 88p.

Taylor, J. C. M., 1977, Sandstones as reservoir rocks, in G. D. Hobson (ed.), *Developments in Petroleum Geology*, 1, Applied Science Publishers, London, 335p.

Telford, W. M., L. P. Geldart, R. E. Sheriff, and D. A. Keys, 1976, *Applied Geophysics*, Cambridge University Press, Cambridge, 860p.

Map Varieties

Maps are a convenient and widely used means of communicating observations, inferences, and conclusions concerning often complex geological relationships. This chapter briefly reviews some of the most widely used map varieties and presents an expanded discussion on contour maps.

GEOLOGICAL MAPS

The best known geological maps (*sensu stricto)* are compilations on various scales of the distribution of rocks and sediments as they occur on the earth's surface. Data are obtained from fieldwork. Rock units (such as formations) or time-rock units (such as series and stages) may be depicted. Most show the distribution of the units as *inferred* from exposures that are variously incomplete. Control may be very good (extensive exposure) or very poor (extensive cover); the extent of control should be shown by "outcrop" overlays or some other means, but rarely is.

A *paleogeologic,* or *subcrop, map* shows the present distribution of rocks below a specific reference surface. The reference surface should be a time plane, whereupon the depiction is of a geological map at that earlier time. Isochronous surfaces can rarely be traced, however, and the practical reference surfaces that are selected (such as unconformities) impart constraints on the meaning of the subcrop map. Such maps are useful for locating truncations and pinchouts of stratigraphic units in which petroleum may be trapped.

Rock distributions above a reference surface (ideally isochronous) are depicted in *worm's eye,* or *lapout, maps,* as viewed from below. Overlap relationships of younger units will be apparent and are useful in prospecting for stratigraphic traps.

Palinspastic maps attempt to show the distribution of rock units before structural deformation (for example, removing the strike-slip movement along the San Andreas or Alpine faults). Such maps are difficult to construct and are subject to much personal bias.

PALEOGEOGRAPHIC MAPS

There are rarely sufficient data to construct detailed paleotopographic maps. More commonly, paleogeographic maps depict gross relationships between land and sea or plains and highlands at specified times in the past. Problems are particularly great in areas where rocks of a particular age are absent—their absence may reflect nondeposition during the time interval (which may be suggestive of terrestrial conditions, for example) or subsequent erosion of rocks once present.

GEOLOGICAL CROSS-SECTIONS

Almost as well known as geological maps are maps that depict the distribution of rocks and sediments in a vertical section through some part of the crust. The line of section is shown on a plan-view geological map; it may be a single straight line or a connected set of straight lines oriented variously. Many cross-sections are based on geological *inference* of the subsurface distribution of the rocks depicted on plan-view geological maps— hence they are second-order interpretations, since those plan-view maps also reflect inferences. Others are constructed by interpreting the distribution of rocks laterally away from vertical control lines that consist of boreholes in which the extent of the

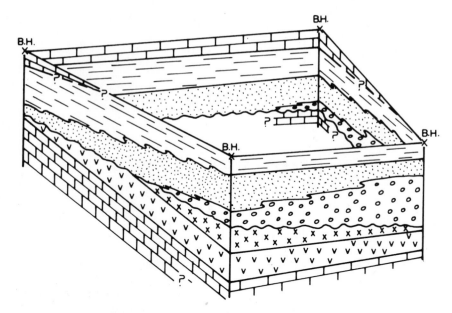

Figure 109. Example of a *panel* diagram.

various units has been measured. A common example of this type of cross-section is the *fence* or *panel* diagram; see Fig. 109.

CONTOUR MAPS

Contour maps show the configuration of real or imaginary surfaces. Any data that can be related consistently to a reference datum and be quantified are mappable. Examples are depths or elevations to a rock boundary from sea level (or an unconformity, or any other reference datum) and the percentage of a particular mineral or rock type (the reference datum being in this case imaginary, i.e., "nothing"—the absence of this material).

A contour line passes through all points of equal numerical value. On maps it is the vertical projection on a piece of paper of the line of intersection between the surface (defined by the data) and an imaginary parallel plane spaced a stated numerical "distance" from the reference datum—for example, a −150-m structural contour is the projected line formed by the intersection of a bedding surface and an imaginary plane 150 m below mean sea level (reference datum = 0 elevation). The distance between contour lines in plan view represents the slope of the surface (hence with structural contours, the dip of the surface may be readily calculated if the horizontal scale is known). The course of the contour is an expression of the irregularity of the surface being mapped (for example, the contours of a vertical cone will be equally spaced concentric circles). See Fig. 110.

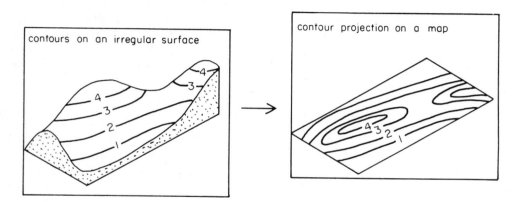

Figure 110. Example of projection for topographic contours.

Contour maps can be used in a number of ways:

1. To determine the relationships between isolated data points.

2. To extend geological inference from areas in which relationships are known into those where control is sparse or absent (that is, to forecast trends).

3. To depict the three-dimensional configuration of rock (or time-rock) units.

4. To provide a means of accurately measuring volumes of irregular masses (such as economic deposits).

5. To determine the relationships between formations, facies, environments, and tectonic or other events.

All contour maps are interpretive. You never *know* the position of all points of a given value; unless data points are closely spaced and the data are accurate, several different patterns are possible with the same data.
Contour patterns can be drawn in three "ideal" ways:

1. By *mechanical spacing*. Where data are accurate and data points are relatively close, the location of contour lines between two points can be found by mechanical interpolation (a contour will pass between the points at a distance from one that is proportional to the difference in data values between the points; see Fig. 111).

2. By *equal slope spacing*. Where data are more inaccurate or data points are relatively widely spaced, a common technique is to *assume* that the surface being mapped has a similar slope throughout the area—that is, contours are drawn with equal spacing throughout the area. The initial spacing should be determined by interpolation between two data points that are relatively close and show the greatest difference in values (where the surface slope appears to be maximum).

3. By *parallel spacing*. Where data are very inaccurate and data points are widely spaced, smooth patterns are constructed by keeping the contours parallel and varying the spacing between them where necessary. Also called *interpretive spacing*, this method can be particularly useful where experience (in nearby or similar areas, for example) suggests that a particular pattern is likely.

In any contouring, all these methods are used in varying proportions. Always try to keep the patterns smooth, parallel, and spaced so as to fill in the whole of the map.
Fig. 112 illustrates the somewhat different patterns that result from several methods applied to two different selection-of-data points from the same set of data—note that resultant maps differ, and see whether you can produce yet other versions by mechanically spacing contours on the two blanks.
Computers can be programmed to construct contour maps, but the results are no more valid than the data fed in and the program used—such maps are not "inherently better" than those constructed by a trained person!

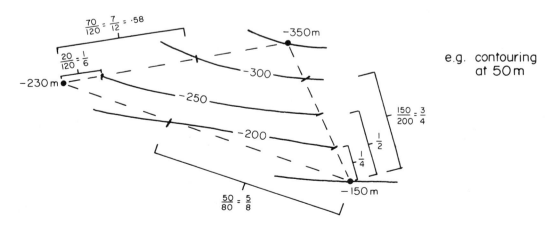

Figure 111. Example of mechanical contouring at 50 m intervals.

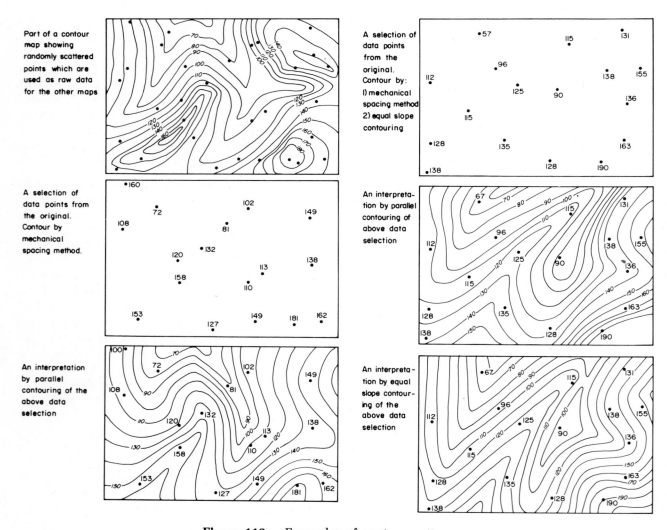

Figure 112. Examples of contour patterns.

SOME RULES FOR CONTOURING

A contour must never end within a map—close it or run it to the map edge.

Where the surface being mapped reverses slope, two contour lines of the same value must appear; that is, a single contour cannot mark the axis of reversal.

The same contour interval should be maintained throughout the map (there are rare exceptions—such as where a log scale is used). The interval selected will be a function of the map scale, the amount of data variation, and the detail required for the mapping purpose.

Contours should not form sharp corners—keep them rounded.

Contours of the same kind should not cross each other (among the exceptions are structural contours of some complex folds—dash or dot the lower contours in the region of overlap).

ADVICE

Don't use ink until the pattern is finalized; draw initial contours lightly and keep an eraser handy.

Begin contouring between data points that show a large difference in value; start a number of contours and take them as a group around the map (rather than trying to trace individual lines).

Break every second, third, or fourth contour line occasionally to insert its value.

Avoid closed contours unless closure is forced by the data.

Always bear in mind the nature of the reference datum and the kind of data you are mapping—it is easy to misinterpret the data until you gain experience.

VARIETIES OF GEOLOGICAL CONTOUR MAPS

Structural contours: where the configuration in space of a real surface is being mapped. A consistently sharp boundary between rock units is necessary. Note that if sea level is the reference datum, large negative data values imply depressions, not highs!

Isopach maps: contours are lines of equal thickness of a rock unit. Consistent sharp upper and lower boundaries to the unit are necessary. For interpretation of sedimentary units, the upper boundary should have been originally horizontal (approximately parallel to sea level); the map then depicts the configuration of the depositional basin unless erosion has subsequently removed part of the unit. The zero contour represents the depositional limit to the unit, or the edge of the unit as influenced by later erosion. The contour pattern may suggest which of these alternatives applies (erosional limits commonly are associated with more rapid thinning of the unit, for instance), but commonly you will require other information (such as lithofacies maps). Isopach maps of a number of stratigraphic units (vertically superimposed or even overlapping) will show basinal evolution. They are used in conjunction with structural contour maps (which show postdepositional deformation), and data points for isopachs may be obtained by simple calculations wherever structural contours on the upper and lower boundaries of the unit intersect. (Note, however, that in these cases the isopach map is a *third*-order interpretation!)

Lithofacies maps: a variety of lithological data can be contoured—for example: percentage of minerals or rock types in a stratigraphic interval (such as percentage of detritals and percentage of sand; ratios, (such as detritals to nondetritals, or sandstones to mudstones); textural maps; and isolith maps, where each contour (isolith) represents an equal thickness of a particular lithotype from within a stratigraphic unit that contains diverse interlayered lithotypes. Ingenuity can produce other varieties, and combinations are possible using colors and ornamentation.

Geophysical contour maps: of gravity or magnetic anomalies, seismic reflection or refraction data, electrical self-potential or resistivity data, geothermal conductivity, radioactivity, and so on.

Geochemical contour maps and others.

Look at varieties of maps in some of the literature referenced at the end of this chapter.

AN EXERCISE ON CONTOUR EXPRESSION OF SUBSURFACE GEOLOGICAL DATA (after LeRoy and Low, 1954)

1. Using a contour interval of 100 ft, from the data given in Table 30, construct the following maps from the blanks in Fig. 113:

(a) Construct a structural contour map of the top of X.

(b) Construct an isopach map of interval X-Y.

(c) Construct a structural contour map of the top of Y.

(d) Test the consistency of your interpretations by selecting a variety of new points on the isopach map and see whether the thicknesses at those locations match the differences between the values at the same locations on the two structural contour maps.

2. Using the data given in Table 31, construct the following maps from the blanks in Fig. 114):

(a) Construct an isopach map (contour interval 100 ft).

(b) Construct a sandstone isolith map (contour interval 10 ft).

(c) Construct a shale and silt isolith map (contour interval 100 ft).

(d) Construct a carbonate isolith map.

(e) Construct a clastic percent map.

(f) Construct an evaporite isolith map (contour interval 50).

(g) Test the consistency of your interpretations by selecting various new points on the isopach map and discover if the thicknesses at these points equal the sum of the isolith thicknesses at the same locations.

3. Discuss the interpretive value of those maps.

Table 30. Exercise on Contour Expression of Geological Data (after L. W. LeRoy and J. W. Low, 1954)

Well No.	Surface Elevation	depths		Datum Elevations		Interval Between X and Y
		X	Y	X	Y	
1	1825	1030	2160	795	−335	1130
2	1835	835	2105	1000	−270	1270
3	1860	790	1970	1070	−110	1180
4	1890	670	2020	1220	−130	1350
5	1845	815	2285	1030	−440	1470
6	1820	720	2250	1100	−430	1530
7	1865	745	2215	1120	−350	1470
8	1840	950	2640	890	−800	1690
9	1835	840	2465	995	−630	1625
10	1845	840	2570	1005	−725	1730
11	1875	995	2685	880	−810	1690
12	1860	1100	3035	760	−1175	1935

Figure 113 *(at right).* Blanks for constructing maps in exercise 1.

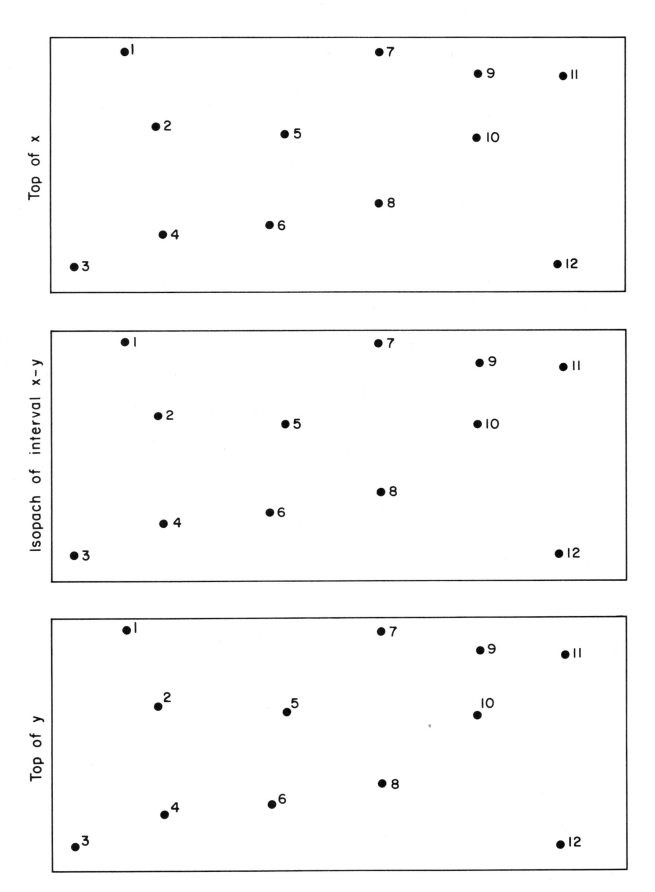

Table 31

Well No.	Total Thickness	Aggregate Thicknesses					Clastic Percent
		Ss.	Sh. and Silt	Ls.	Dolo.	Evapor.	
1	0	0	0	0	0	0	0
2	570	0	230	102	238	0	40
3	580	0	90	172	318	0	15
4	775	0	110	98	552	15	14
5	90	0	90	—	—	0	100
6	430	Trace	400	30	0	0	93
7	130	0	130	—	—	0	100
8	0	0	0	—	—	0	0
9	860	0	140	130	520	70	16
10	0	0	0	0	0	0	0
11	540	0	390	135	15	0	72
12	420	25	395	0	0	0	100
13	530	10	500	15	5	0	96
14	280	18	222	18	22	0	83
15	620	8	320	132	160	0	52
16	0	0	0	0	0	0	0
17	0	0	0	0	0	0	0
18	510	22	200	149	121	18	43
19	85	10	75	—	—	0	100
20	210	12	198	—	—	0	100

Figure 114. Blanks for constructing maps in exercise 2.

Isopach Map

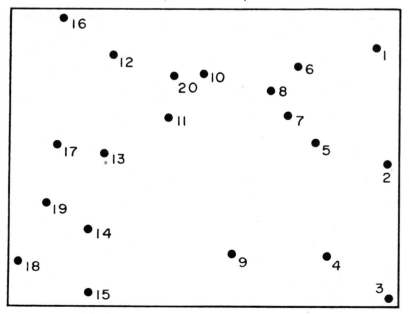

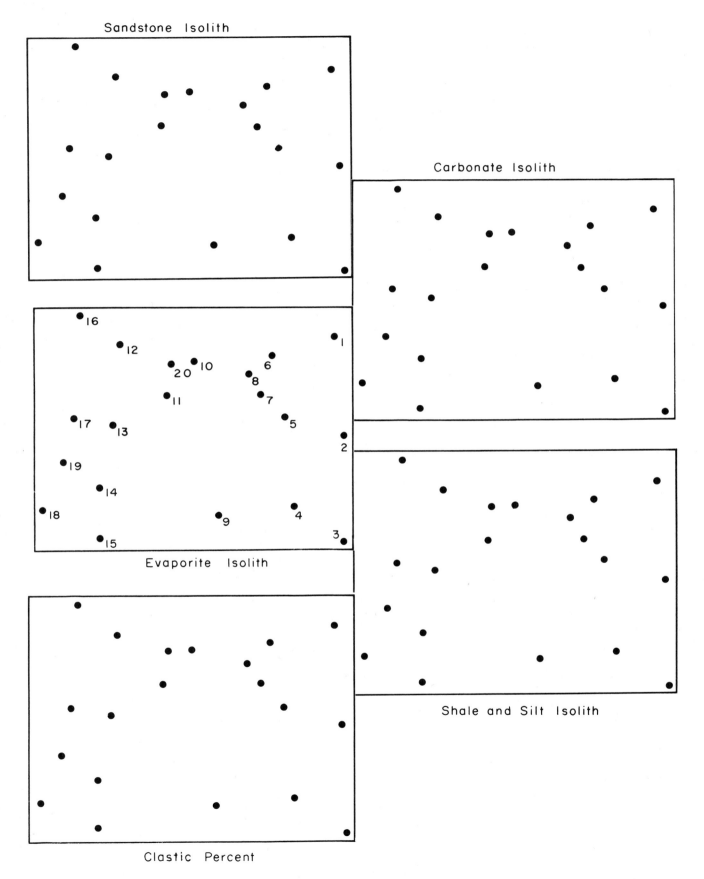

Sandstone Isolith

Carbonate Isolith

Evaporite Isolith

Shale and Silt Isolith

Clastic Percent

LITERATURE

Bishop, M. S., 1960, *Subsurface Mapping*, Wiley, New York, 198p.

Busch, D. A., 1974, *Stratigraphic Traps in Sandstones—Exploration Techniques, Am. Assoc. Petroleum Geologists Mem.* 21, 174p.

Irvine, J. A., S. H. Whitaker, and P. L. Broughton, 1978, *Coal Resources of Southern Saskatchewan: A Model for Exploration Methodology*, Geol. Survey Canada Econ. Geology Rep. 30, Geological Survey of Canada, Ottawa, 151p. and 56 plates.

Kay, M., 1945, Paleogeographic and palinspastic maps, *Am. Assoc. Petroleum Geologists Bull.* **29:**426-450.

Krumbein, W. C., 1952, Principles of facies map interpretation, *Jour. Sed. Petrology* 22:200-211.

Krumbein, W. C., 1955, Composite end members in facies mapping, *Jour. Sed. Petrology* **25:**115-122.

Krumbein, W. C., 1959, Trend surface analysis of contour-type maps with irregular control-point spacing, *Jour. Geophys. Research* 64:823-834.

Le Roy, L. W. (ed.), 1950, *Subsurface Geologic Methods—A Symposium*, 2nd ed., Colorado School of Mines, Golden, 1166p.

Le Roy, L. W., and J. W. Low, 1954, *Graphic Problems in Petroleum Geology*, Harper & Brothers, New York, 238p.

Levorsen, A. I., 1960, *Paleogeologic Maps*, Freeman, San Francisco, 174p.

Low, J. W., 1958, Subsurface maps and illustrations, in J. D. Haun and L. W. Le Roy (eds.), *Subsurface Geology in Petroleum Exploration*, Colorado School of Mines, Golden, 453-530.

McKee, E. D. (ed.), 1956, Paleotectonic maps—Jurassic System, *U.S. Geol. Survey Misc. Geol. Inv. Maps* 1-175.

McKee, E. D. (ed.), 1959, Paleotectonic maps of the Triassic System, *U.S. Geol. Survey Misc. Geol. Inv. Maps* 1-300.

Moody, G. B. (ed.), 1961, *Petroleum Exploration Handbook*, McGraw-Hill, New York.

Pelto, C. R., 1954, Mapping of multicomponent systems, *Jour. Geology* **62:**501-511.

Rees, F. B., 1972, Methods of mapping and illustrating stratigraphic traps, in R. E. King (ed.), *Stratigraphic Oil and Gas Fields, Am. Assoc. Petroleum Geologists Mem.* 16,168-221.

Schuchert, C., 1955, *Atlas of Paleogeographic Maps of North America*, Wiley, New York, 177p.

Sebring, L. Jr., 1958, Chief tool of the petroleum geologist: the subsurface map, *Am. Assoc. Petroleum Geologists Bull.* **42:**561-587.

Suggestions for Fieldwork

In the field it is vital to establish a systematic pattern of work—for field observations, for note-taking, for sample collecting, and for evenings and rainy days. Plan the daily operations in the framework of the project as a whole—time is wasted on indefinite traverses. Plot your route on both your base map and air photos. Following an orderly sequence of observations and notation—for example, structure then texture then composition—reduces the time you spend per exposure and establishes a routine for comprehensive description; you do not want to discover later a need to return to an exposure for details missed the first time. Table 32 provides an example of a field checklist. Index your notebook so that you *and* others can easily and rapidly find data, and cross-index (with systematic locality numbers) to any samples, photographs, and the map and/or air photos. During evenings and on (impossibly) wet days, update and interpret your map, compile detailed sample descriptions, and plan future field and laboratory operations. The following discussion only briefly introduces generalities for fieldwork—consult some of the literature at the end of this chapter, such as Lahee (1961) for comprehensive discussion and detailed techniques and procedures.

FIELD MAP

Put as much information as possible on the map or photo overlay in the field with a hard, sharp pencil. Outline (or mark with an "x") all exposures. Sketch in contacts, faults, and dikes; show strike and dip wherever measured, sample localities, extent of exposure, and so on. Do *not* rely on your memory to insert these data later! Use solid lines for contacts, faults, and the like that occur unequivocally within the area represented by a line width on the map or air photo overlay (on small-scale maps or photos, a line width may represent a band 10 m wide!). Use dashed or dotted lines where uncertainty exists. Make notations on the extent of "cover" so that you realize at a later date that you have traversed the area.

Locate reference points (such as sample localities) as precisely as possible. Depending on the scale of your field sheet and the locality (say, dense bush), it may be necessary to take a bearing on an easily identified reference point and pace the distance to your locality (or triangulate by compass). When taking bearings on hilltops, be sure that you can see the true top from your locality. If you can find a distinctive feature that is at your own ele-

---

Table 32. Checklist for Field Observations

Preliminary Notes

Number pages in upper-right hand corners; leave left-hand or top pages for sketches.

General title and location of project.

Field Party (note each day if party varies).

(Continues on next page)

215

Table 32. *(Continued)*

EACH DAY

Date; weather conditions

Map and/or aerial photograph number (s).

EACH LOCALITY: LOOK OVER THE ENTIRE EXPOSURE FIRST

Locality number (consecutive within project).

Exact location (grid intersection from map or show locality no. on field sheet), AND brief notation of physiographic setting—e.g., stream cut in heavy bush.

Exposure character (size, shape, and quality of exposure).

Rock unit name (not necessarily *formal* name).

Approximate stratigraphic position within major unit (if possible).

Approximate stratigraphic thickness of each unit exposed.

Description of boundaries of each unit recognized and spatial relationships with other units exposed.

Structural Data

Attitude of principal bedding planes, variability in strike and/or dip. For strike and dip, use consistently either azimuth (0-360°) or specified quadrant (e.g., S45E); specify direction of dip (e.g., 20SW).

Nature of stratification—scale and character of irregularities, sharp or gradational contacts, thickness of strata.

Other primary structures—e.g., graded bedding, cross-bedding, channels, ripple-mark, flute marks, concretions, mud cracks etc. Note characteristics of each—scale, extent of development, attitude etc. Measure orientation or direction of paleocurrents.

Facing-direction (which way was up?); what evidence, how convincing?

Folds—sketch and annotate. Symmetry, type of fold form, bearing and plunge of axial plane trace, and trace of the crest or trough. Relationships to other structural features.

Faults—sketch and annotate. Altitude, extent and character of fault zone, direction of movement (apparent vs. real?), bearing and plunge of any striations.

Cleavage (and foliation, schistosity) (S_1, S_2, S_3, etc in order of relative age; SO = sedimentary bedding)—attitude and characteristics, relation to lithology and other structures.

Joints (G_1, G_2, etc.)—attitude, degree of development.

Lineation—note if primary current lineation (L_0).L_1 = intersection of S_0 and S_1, L_2 = intersection of S_1 and S_2, etc. Preferred mineral orientation in igneous/metamorphic rocks. Take bearing and plunge or pitch.

Flow banding in igneous rocks; presence or absence of baking at contacts with sediments; chilled margins; cross-cutting relationships.

Table 32. (Continued)

Description of Lithology (begin at the base of any unit and work upward)

Graphic Log?

Color (wet vs. dry)—use discrete colors (not buff, chocolate, etc), preferably use color chart for comparison.

Weathering—degree and depth.

Permeability/porosity—qualitative observation of obvious holes or rapidity of water absorption.

Induration—degree; is it due to any obvious cement?

Texture—size (max., min., mode, proportions), sorting, shape and roundness, fabric, quantity of matrix, matrix- or clast-supported? Textural classification. With igneous rocks, note size range of phenocrysts and fabric.

Composition—identify components as possible, estimate percent of each. With conglomerates, note abundance of pebbles of same (specified) lithology. Note any relationship between composition and texture. Compositional classification. For metamorphic rocks, relate mineralogy to structure.

Fossils—although part of the "composition" of a rock, detailed notes are often useful on their occurrence (e.g., *in situ*?) and character (identify to level possible, collect unknown or potentially useful samples), the relative abundance of each type, apparent associations, morphology of odd types. Sketch. Are they (generally or as individual species) uniformly distributed or concentrated in layers, lenses, concretions? Do they show a preferred orientation (e.g., individual valves concave upward)? Proportion of articulated forms? Degree of abrasion/fragmentation? Further data should be collected when paleoecological study is warranted (special checklists necessary).

Trace fossils—describe toponomically, noting size, morphology, and relationship to sedimentary structures and body fossils. Give formal genus/species name if known. Infer ethological class if warranted. Note any difference in sediment texture/composition within the trace relative to that surrounding it. Check for evidence of boring vs. burrowing origin where relevant. Note concentrations of traces relative to litholoy and sedimentary structures.

Relationships of Units

Physiographic—whatever seems useful for later report. E.g., "This unit forms a waterfall across the stream".

Stratigraphic—whatever seems appropriate without being repetitive. E.g., "similar units appear to be present 50 m above and below" or "unlike the same unit to the west in having ...."

Vegetation—do certain flora grow on, or avoid, the unit in preference to other lithologies?

Soil—any apparent relation to bedrock?

Interpretations

Jotting down ideas on origin and possible interrelationships are well worthwhile when they come to you. But they will generally change later, so keep them confined to a separate section in your notes rather than juxtaposed with observations.

vation (by sighting along a leveled compass), you can find your location easily on a topographic map. Initially, do not waste time in areas where accurate location is difficult—for example, leave broad bushed areas to be examined when you have a reasonable idea of what you'll find therein.

Use a consistent *standard* set of symbols (see Fig. 115) and abbreviations (such as in Table 33), both for yourself and for other potential users of your map or notebook; it is wise to list these on the first or last page of your notebook.

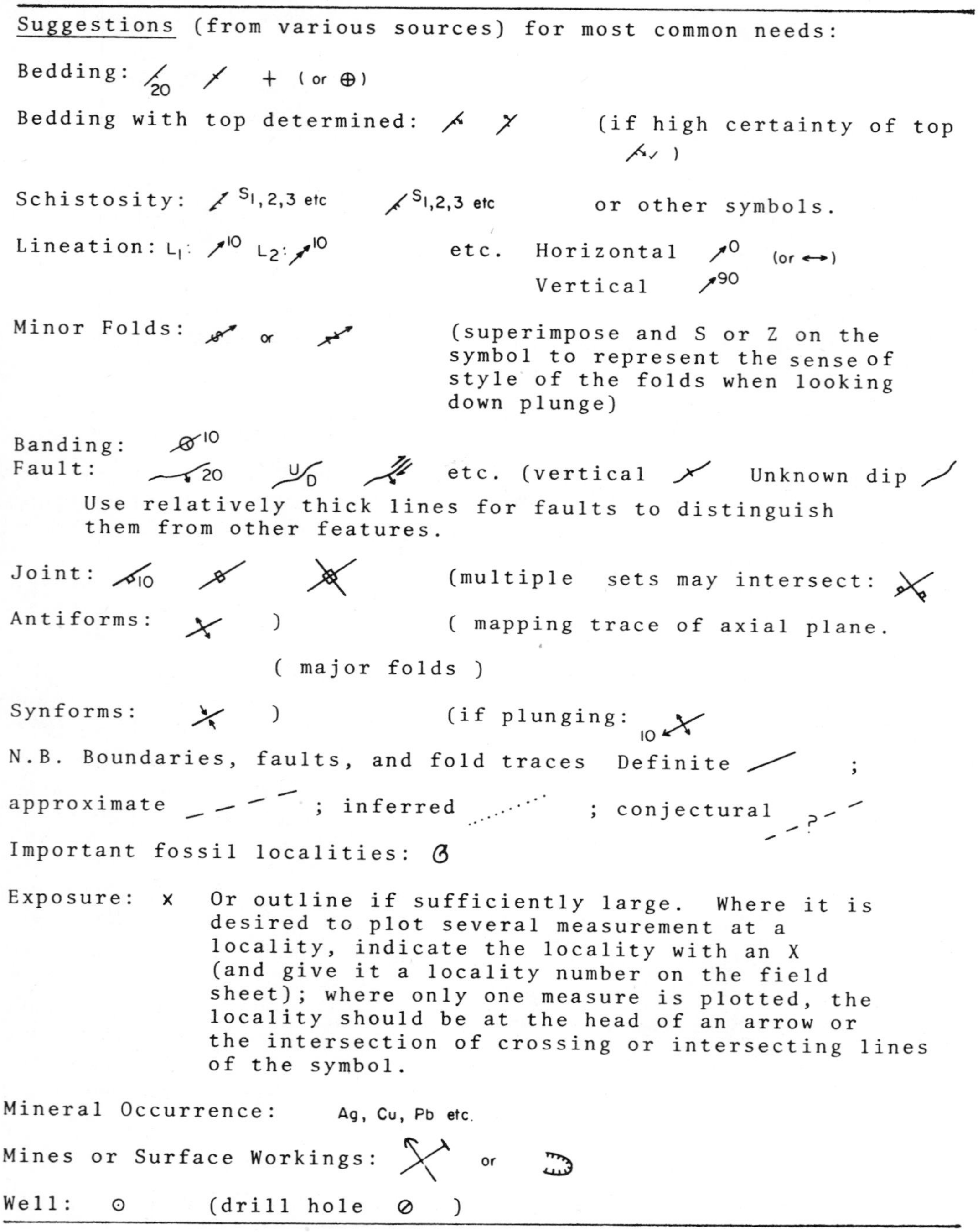

Suggestions (from various sources) for most common needs:

Bedding: /20 / + (or ⊕)

Bedding with top determined: / /   (if high certainty of top /⌄)

Schistosity: / S1,2,3 etc   / S1,2,3 etc   or other symbols.

Lineation: L1: /10 L2: /10   etc. Horizontal /0 (or ↔)
                                          Vertical /90

Minor Folds: ⌐ or ⌐   (superimpose and S or Z on the symbol to represent the sense of style of the folds when looking down plunge)

Banding: ⊘10

Fault: —20   U/D   etc. (vertical ✗   Unknown dip /)
    Use relatively thick lines for faults to distinguish them from other features.

Joint: /10   ✗   ✗   (multiple sets may intersect: ✗)

Antiforms: ✗)   (mapping trace of axial plane.
          (major folds)

Synforms: ✗)   (if plunging: 10 ✗)

N.B. Boundaries, faults, and fold traces Definite / ;

approximate _ _ _ ; inferred .... ; conjectural _ _? _

Important fossil localities: ℬ

Exposure: ✗ Or outline if sufficiently large. Where it is desired to plot several measurement at a locality, indicate the locality with an X (and give it a locality number on the field sheet); where only one measure is plotted, the locality should be at the head of an arrow or the intersection of crossing or intersecting lines of the symbol.

Mineral Occurrence: Ag, Cu, Pb etc.

Mines or Surface Workings: ✗ or ⋙

Well: ⊙   (drill hole ⊘)

Figure 115.   Map symbols. Every organization will have its own standardized map symbols. Be sure you have a list of their standards with you when you are in the field, and use them from the first to avoid confusion and revision in preparation of the final map from your field sheet.

Table 33. A Selection of Useful Abbreviations

bx	breccia	lt	light
cgl	conglomerate	dk	dark
Ss	sandstone	bn	brown
Zs (sltst)	siltstone	gy	grey
Ms	mudstone	sil	siliceous
dol	dolomite, dolostone		
ls (lmst)	limestone	bdg	bedding
gn	gneiss (or green)	fm	formation
sch	schist	mb	member
rk, rx	rock, rocks	jt	joint
calc	calcareous	met	metamorphosed, metamorphic
ferrug	ferruginous	ig (n)	igneous
qtz	quartz	sed	sedimentary, sediment
xl (n)	crystal (-line)		
		spl	sample
tr	present in trace amounts ($< 1\%$)	exp	exposure, exposed
v	very	ph	photo *(number)*
avg	average		

PLANNING TRAVERSES

Initial traverse: Make your first observations along streams and roads. These are not only easy to locate on the map, but usually provide good exposures of rock. Next in ease of location are fence lines and power lines, although they have the disadvantage of deflecting the compass. Sharp ridges are ideal for initial traverses because you can keep landmarks in sight and locate yourself easily on a contour map, but on a photo their location is not easy.

Cross-trend traverses: The initial traverses will tell you the general distribution of rock types in the area; with this information, you may plan a series of traverses—say, 1000 m apart—back and forth across the "grain" of the geology, always keeping yourself accurately located. The cross-trend traverses will solve some problems, and where the geology is simple it may be possible to match contacts and run them across from one traverse to another by a careful study of an air photo and the topography.

Walking contacts: Where cross-trend traverses do not match, or the geology is complex, it becomes necessary to map the intervening area by carefully "walking out" selected contacts. Suppose that formation A is on the downhill side, and on your left. Trace it by keeping the uppermost fragments (float) of A on your left and the lowermost outcrops of B

on your right, mapping and recording as you go. If you lose the contact entirely, start again at its next appearance and work back into the doubtful area.

USE OF FLOAT

Float consists of angular rock fragments on a slope, derived from nearby bedrock by mass wasting. Rarely are more than two or three rock types found together; the physiographically lower rock units can be progressively mapped along uphill traverses according to the progressive disappearance of each type of rock fragment. Draw contacts as approximate.

Debris transported by streams is usually lithologically diverse, but progressive upstream disappearance of particular rock types may indicate where to look for contacts (for example, some tributaries may drain areas of distinctive bedrock).

Soil may be mapped as bedrock where it is clearly residual (derived from the bedrock). When mapping Quarternary geology, a transported soil may be one of your map units!

TECTONIC FEATURES

Large faults and folds usually become clear only as a result of detailed mapping of contacts and the synthesis of numerous dips and strikes on a complete map. Do *not* expect to see them in the field.

Small faults and folds may be very useful in determining the tectonic history, but they can be confusing if their origin is uncertain.

Minor folds and faults of tectonic origin are produced by the same regional stresses that produced the major structures, hence they are usually consistent within the same exposure and show a similar pattern to those found nearby. If faults are too small to map individually, nevertheless note the strike, dip, and direction of movement in your notebook—the data may be useful during general synthesis. With minor folds and crumples, map the axial trends and their average plunge.

Sedimentary (penecontemporaneous) folds and faults result from gravitational slumping and deformation before lithification. Most are confined to a few beds in a sequence and they are often irregular and/or inconsistent, showing no relation to the regional tectonic pattern. Diffuse contacts between strata or other features may suggest the nonlithified character of the sediments during deformation.

Fracturing and/or distortion by mass wasting processes may result from modern weathering, soil formation, tree growth, or downslope creep. Observation will generally reveal a haphazard pattern or a direct relationship to the topography. Do not map these features except for special purposes.

SELECTION OF GEOLOGICAL UNITS

In an area of arenites and limestones, it is obvious that the arenites must be mapped separately from the limestones, but how many rock units of each type should be distinguished? Just as every sample of granite tends to differ in detail, so each sample or exposure of sediment may have different characteristics. Differences apparent in the field may be greater within one limestone unit than between two limestones separated stratigraphically by an intervening arenite. Here are some general suggestions for selecting and recognizing mappable units:

1. The larger the scale of the map and the more time you have available, the more subdivisions you should attempt. A single continuous set of strata may be subdivided into members in the notes at any one locality, even if the members are too small to map there; further work in the region will ultimately show whether such members should be extended and formalized. It is better to subdivide too much in the beginning; you can drop divisions later if they prove useless.

2. Distinguish rock units on the basis of characteristics that are likely to persist within that unit. Sedimentary structures may vary greatly; composition is most likely to remain consistent. (Unfortunately, composition in particular may be similar in stratigraphically separate units.) Color is not always a valid criterion—oxidation or reduction, for example, may reflect localized diagenetic or weathering effects; fresh and weathered rocks may look quite different.

3. The nature of contacts or the recognition of stratigraphic sequence generally may prove useful (for example, perhaps there will be distinctive units above or below the problematical one).

4. Where stratigraphic position within a formation is difficult to determine in the field because of irregular or apparently regular alternation of lithologies (such as interbedded sandstones and mudrocks), try measuring the range of bed thickness and texture in each exposure. When you finally review your compilation of data, you may find evidence of vertical or lateral changes within the formation as a whole within your area, and these changes (such as in the ratios of sandstone to mudstone) may be sufficient to warrant subdivision, even if on the basis of approximate or arbitrary boundaries.

5. Do not be concerned with the formal naming of the rock units in the early stages of mapping (unless you recognize distinctive units from adjacent areas already mapped). Use purely descriptive names of your own choice (such as "Third Arenite," "Mottled Limestone") and decide on final names later.

AGE OF UNITS

Assigning a geological age to a stratigraphic unit is not particularly important in mapping operations within a local area, and it is generally not feasible unless you find and recognize diagnostic fossils. It is better merely to assign relative ages determined by the principles of stratigraphic superposition and await laboratory-based age determinations than to guess from similarities with other areas.

METAMORPHIC FEATURES

In a regionally metamorphosed terrain, the lithological units present will probably have been complexly folded several times. Although it is sometimes possible to elucidate much of the major structure using minor fold and lineation interrelationships, it is advisable to establish whether a characteristic lithological succession or some mappable marker horizon is present; always start mapping from the region where this succession or horizon is clearly seen. To achieve any satisfactory metamorphic history, you must map in great detail. Record every minor fold and lineation, preferably on a stereographic projection. It is important to establish which lineations or folds are refolded or superimposed on other lineations and folds. The history of mineral growth and recrystallization can be related to structures in the field (but should also be studied in thin section). Ultimately, the development of minor structures and mineral growth should be related to major structures on the basis of field mapping of the lithological succession.

MEASUREMENT OF GEOLOGICAL SECTIONS

Locations of stratigraphic sections for detailed measurement are best selected after areal mapping is almost completed. You must evaluate accessibility as well as the quality and extent of exposure.

Where beds dip over 10° and there is little relief along the line of section, pace-and-compass measurement is feasible if distances are great and lithological character does not vary greatly over short intervals. Measurement by tape traverse is necessary where variation is rapid and highly detailed description necessary.

Where beds dip over 10° and there is marked relief along the line of traverse, plane-table surveying and graphical solution for dips are commonly necessary, unless a very accurate, large-scale topographic map is available.

Where beds are nearly flat-lying, using a hand-level or Brunton clinometer while pacing is reasonable if distances are not too great.

Because the results of section measurement will generally be presented in the final report as a columnar section, it is wise to attempt direct recording of data on graphical logs (for example, see Bouma 1962; Andrews, 1982). An example of the layout of one such log is provided (Fig. 116), and suggested symbols shown on Fig. 117.

SAMPLING

Samples are collected for refined analysis in the laboratory, and for comparison with other samples to help identify mapping units in the field. Show sample numbers on your field map at the sampling localities, and note details of the relationship of the sample to the exposure in your notebook (for example, by a sketch). Ensure that samples are adequately and indelibly labeled and located while you are in the field—do not rely on your memory! A sample numbering system can be devised that will provide considerable information itself—for example, "BL-18-81-2" might mean "second sample collected from locality 18 during 1981 by the field party of Bradshaw and Lewis." An ƒ could be added to specify fossil locality. Personal initials are useful in the notation if there is a possibility of mixing samples from different field parties; the year or some other index can help avoid confusing samples collected by the same person on different jobs.

A *representative* sample of each new rock type and fossil is generally necessary when you make a detailed study of an area. Samples from widely separated (vertically or laterally) positions in the same rock unit are worth taking for comparison—laboratory analysis may show differences indistinguishable in the field. Clearly note when you have selected samples showing unusual characteristics.

Samples should be fresh; an optimal size is approximately 10 × 8 × 5 cm.

BASE WORK—EVENINGS AND IMPOSSIBLE WEATHER

1. Go over all observations on the field map or air photo overlay with a very fine pen using India ink (such as an "O" Rapidograph). Do not extend contacts in ink beyond the extent observed in the field. Light coloring-in of distinctive lithologies may help you visualize relationships and future fieldwork. *Keep the field map up to date.*

2. Transfer data from your field sheets to the office map—put raw data in ink as on the field sheets. Using a hard, sharp pencil and without pressing too hard, extend contacts and extrapolate data as you deem useful for interpreting relationships and planning future fieldwork (this will help you pinpoint problem areas, for instance). This step is particularly essential when you are working with a number of air photos, but you should not neglect it in any case—the office map is the basis of your final map, and by the time you leave the field it should include virtually all the data that you could possibly need for the final map.

Locality:

Date:

Worker:

Sheet No:

Total Thickness

Thickness

Outcrop Profile
Rock Type
Bedding Contact
Rel. Induration

Sedimentary Structures
Bedding Plane (Top) (Base)
Internal

Current Direction

Colour

Texture
Gravel (specify)
Sand — VC C M F VF
Mud
C(lay) Si(lt) predom.

Composition

Fossils

Remarks (including Composition) and/or sketches

S(ample) P(hoto) No.

Figure 116. Example of graphic log for use in the field.

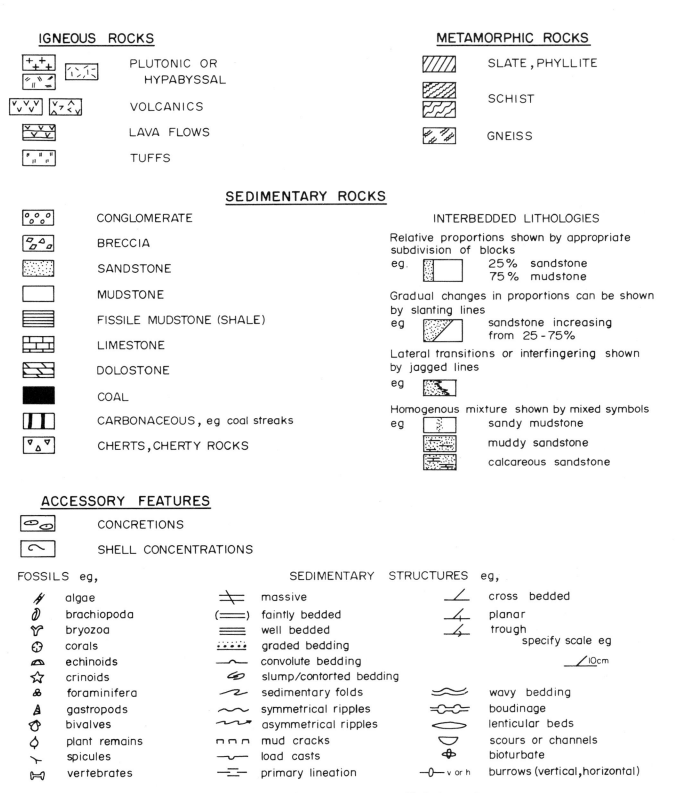

Figure 117. Selected patterns for representation of lithologic characteristics on cross-sections, graphic logs, etc.

3. Organize your data from the field notebook. Write descriptions on separate pages for each lithological unit so that the dispersed field observations are together. Review notes to ensure that all data you will need is there; decide what observations are critical when you next encounter each unit in the field. Think about interpretations and seek further data to substantiate or disprove them. When you come to write your final report, the data and some of your interpretations will be already organized.

4. Compare all samples from the same units. Examine each sample in detail for properties you may have missed in the field. Select those requiring laboratory analysis.

5. Plan your next day's fieldwork. Examine your map and air photos, mark probable exposures, plot your intended traverse, and so on. Plan to spend maximum time in the problem areas pinpointed by the preceding review.

FINAL REPORT

The preparation of a final report may take as long as, or longer than, the fieldwork itself, depending on the amount of laboratory and library work necessary. The following suggestions apply to most final reports; your instructor, and later your employer, will indicate the requirements for a specific report. Write at least one rough draft; let it lie for some time, then reread, correct, shorten, and rewrite. Give it to a colleague to review; evaluate his or her comments and revise accordingly. Proofread.

Be concise: make full use of diagrams, photos, and tables to conserve words. Organize carefully; rampant disorganization is also a widespread disease. Type (or write or print legibly) on one side of the paper only.

It is generally a good idea to compose a super-detailed table of contents before writing; it will usually be revised several times during the course of writing, but initially it will help you organize your thoughts as well as the report. Use plenty of headings and subheadings in the report to aid clarity and help readers find their way around. Table 34 provides an example of organization.

Be consistent—describe the same kinds of features in the same order for each rock unit; develop discussions in a parallel manner for clarity and ease of comparison. Beware of internal contradictions (both in the text and between the text and the map).

Some useful guides to writing are provided in the following literature list. Keep a dictionary, a glossary, and a thesaurus at hand.

LITERATURE

Fieldwork

Andrews, P. B., 1982, *Revised Guide to Recording Field Observations in Sedimentary Sequences,* New Zealand Geol. Survey Rep. 102, 74p.

Berkman, D. A., and W. R. Ryall (eds), 1976, *Field Geologists Manual,* Australasian Institute of Mining and Metallurgy, Parkville, Australia, 295p.

Bouma, A. H., 1962, *Sedimentology of Some Flysch Deposits,* Elsevier, Amsterdam, 458p.

Compton, R. R., 1962, *Manual of Field Geology,* Wiley, New York, 378p.

Kottlowski, F. E., 1965, *Measuring Stragigraphic Sections,* Holt Rinehart and Winston, New York, 233p.

Kupfer, D. H., 1966, Accuracy in geologic maps, *Geotimes* **10**(7):11-14.

Lahee, F. H., 1961, *Field Geology,* 6th ed., McGraw-Hill, New York, 926p.

Low, J. W., 1957, *Geologic Field Methods,* Harper, New York, 489p.

Tucker, M. E., 1981, *The Field Description of Sedimentary Rocks,* Halsted Press, New York, 128p.

Communication

American Association of Petroleum Geologists (AAPG), 1970 and later eds. *Slide Manual,* Tulsa, Okla., 32p.

Blackadar, R. G., 1968, *Guide for the Preparation of Geological Maps and Reports,* Geol. Survey of Canada Misc. Report 16, Department of Energy, Mines and Resources, Queen's Printer, Ottawa, 147p.

Bryant, D. C., and K. R. Wallace, 1976, *Oral Communication,* 4th ed. Prentice-Hall, Englewood Cliffs, N.J., 270p.

Clanchy, J., and B. Ballard, 1981, *Essay Writing for Students,* Longman Cheshire, Melbourne, 124p.

Clifton, H. E., 1978, Tips on talks and how to keep an audience attentive, alert, and around for the conclusions at a scientific meeting, *Jour. Sed. Petrology* **48**:1-5.

Cochran, W., P. Fenner, and M. Hill (eds.), 1974, *Geowriting,* American Geological Institute, Washington, D.C., 80p.

Crutchley, B., 1970, *Preparation of Manuscripts and Corrections of Proofs,* 6th ed., Cambridge University Press, Cambridge, 19p.

Day, R. A., 1979, *How to Write and Publish a Scientific Paper,* ISI Press, Philadelphia, 160p.

Fowler, H. W., 1965, *A Dictionary of Modern English Usage,* revised by E. Gowers, Oxford University Press, Oxford, 725p.

Table 34. Example of Final Report Organization

Title: Combine brevity with clarity and precision.

Abstract: Single-spaced; not over 150 words. Give salient facts and conclusions only. Write this section after completing the rest of the report.

Table of contents: Headings and subheadings in the same form as in the text; page numbers where discussion begins in text.

List of figures: Includes drawings, photographs, and map(s). Titles as in text.

Introduction

  Purpose

  Methods (field party, time spent, maps and aerial photos, methods of study)

  Physiographic Setting (location of area, topography, climate, drainage, vegetation, access, nearest population center(s), size and limits of map area)

  Geological Setting (very brief resumé of broad relationships to major tectonic, time, and widespread or well-known stratigraphic units)

  Previous Work (brief review of work specifically on the area with which you are concerned)

  Terminology (used in present report where it differs from that used by other reputable geologists, e.g., "greywacke" or term used in broader or more restrictive sense than originally defined)

Geomorphology: Description and explanation of all significant landforms (discussion of this section may be more appropriate elsewhere; in some cases it need not be included.)

Stratigraphy: Will have numerous headings and orders of subheadings—be consistent in order and format. Treat intrusive igneous rocks and tectonic structures in sections separate from sedimentary rocks. Treat systematically from oldest to youngest units within each section.

  Stratigraphic Unit (defined; extent; best exposures; thickness; list of recognized subunits)
  (Oldest subunit—as above)
  Primary structures
  Lithology—color, induration, texture, composition; variability within area.
  Paleontology
  Inferences—geologic history (portions or whole; may be deferred)
  Contact with underlying rocks—nature, recognition criteria, rough time span represented.

Igneous rocks (intrusives): An introductory paragraph delineating relationships may be useful.

  Lithology and name (definition, how and why defined; extent; characteristic exposures; broad lithological characteristics; subdivisions)

  (Subdivision) Petrology (include features evident in field—e.g., foliation, jointing, weathering, variation in properties)

(Continues on next page)

Table 34. *(Continued)*

Relation to surrounding rocks (contacts, form of body; relative age)

Petrogenesis—genetic inferences

Structure

Folds; Faults (including resultant features such as schistosity and lineations)— Major vs. minor structures; field evidence, areal extent; form and charac- teristics; age relationships; units involved and not involved; degree to which presence is inferred vs. observed.

Joints—sets, systems; consistency; relation to folds and faults.

Interpretation—causal factors; depth of formation; integration with structure outside of map area.

Quaternary Deposits (to extent warranted by purpose of project)

Economic Geology: Describe mineral deposits, workings, claims, etc. Subdivide as suggested by economic history of region or type of deposits.

Geological History: Synthesis of preceding sections oriented to sequential development of geological features of region. May take the place of *Summary.*

Summary (May precede *Geological History* or take the place of *Conclusions)*

Conclusions (Salient discoveries of project; may be replaced by *Geological History* or *Summary,* depending on project).

Final Map: Clean, neat, inked; show scale, north line, declination, latitude and longitude, reference grid system, legend, title, date, name of geologist. Preferably include an exposure map (overlay) as well as the interpretive map. (See Kupfer, 1966.) Color in units; use reference symbols where useful. Use solid lines where geological features are certain within line width.

Cross Sections (aside from title and scale, no extra legend should be necessary): One or more are generally necessary with the geological map to show the inferred geological relationships. Draw cross-sections through the regions most in need of clarification; the lines of section need not be straight throughout the area. The north or east end of the section should be to the right. Choose the *vertical exaggeration* with care (and do not forget to record it). Do not extrapolate *doubtful contacts*—terminate them with a small question mark.

Columnar Section: A profile columnar section (e.g., see Lahee, 1961) should be drawn for the sedimentary rocks. Age and formation names appear on the left, lithology and fossils on the right. Breaks and interruptions in the record are indicated by jagged breaks in the columns. Intervals of known thickness but unknown lithology (e.g., covered in the field) may be left blank in the unbroken column. With careful preparation, the columnar section may serve as the lithologic legend for the map.

Gowers, E. A., 1957, *The Complete Plain Words*, Her Majesty's Stationery Office, London, 209p.

Hicks, T. G., 1961, *Writing for Engineering and Science*, McGraw-Hill, New York, 298p.

Kirkman, J., 1975, That pernicious passive voice, *Phys. Tech.*, 197–200.

Landes, K. K., 1966, A scrutiny of the abstract, *Am. Assoc. Petroleum Geologists Bull.* **50**:1992. (Also see p. 1993, Guide for preparation and publication of abstracts by The Royal Society.)

Murray, M. W., 1968, Written communication—a substitute for good dialogue, *Am. Assoc. Petroleum Geologists Bull.* **52**:2092–2097.

O'Connor, M., and F. P. Woodford, 1976, *Writing Scientific Papers in English: An ELSE-Ciba Foundation Guide for Authors*, Elsevier/Excerpta Medica/North Holland, Amsterdam, 108p.

Partridge, E., 1973, *Usage and Abusage*, Penguin, Harmondsworth, England, 380p. (reprinted in 1974, 1975, 1976).

Rodolfo, K. S., 1979, One picture is worth more than ten thousand words: how to illustrate a paper for the *Journal of Sedimentary Petrology, Jour. Sed. Petrology* **49**:1053–1060.

Royal Society, London, 1974, *General Notes on the Preparation of Scientific Papers*, rev. ed., Cambridge University Press, Cambridge, 31p.

Shinn, E. A., 1981, Make the last slide first, *Jour. Sed. Petrology* **51**:1–6.

Strunk, W. Jr., and E. B. White, 1972, *The elements of Style*, 2nd ed., Macmillan, New York, 78p.

Turabian, K. L., 1960, *A Manual for Writers of Term Papers, Theses, and Dissertations*, University of Chicago Press, Chicago, 164p.

U. S. Geological Survey, 1958, *Suggestions to Authors of the Reports of the United States Geological Survey*, U. S. Government Printing Office, Washington, D. C., 255p.

Vallins, G. H., 1951, *Good English: How to Write It . . .* Deutsch, London, 255p.

About the Author

DOUGLAS W. LEWIS was born in Brazil, of U.S. parents, and grew up mostly in England and France. He continued his roving by taking the B.A. degree at Cornell University in 1959, the M.Sc. at the University of Houston in 1961, and the Ph.D. at McGill University in 1965 and by traveling widely in the United States and Canada. His wife and one of his daughters are true kiwis, born in New Zealand, whereas the other daughter is a "Connecticut Yankee." Originally intending in 1965 a three- to four-year stint in New Zealand, he became enamoured of the country's geology and finally grew some roots; Dr. Lewis has been at the University of Canterbury since 1965 and is currently head of the geology department. Subsequent geological visits to western, central, and eastern Australia and two study leaves concentrated in North America, but incorporating visits to Bahrain, Iran, Israel, and Scotland, have satisfied his wanderlust to date. His next leave will be based at San Diego State University (during the first half of 1984), where he will be teaching sedimentary geology as part of an exchange with a staff member there.

Dr. Lewis purposefully spread his research interests over a wide range of topics in sedimentary geology to improve his teaching ability. He has published papers on detrital and carbonate rocks, emphasizing aspects as diverse as arenite petrography, primary and secondary structures, glauconite, paleoenvironmental analysis, diagenesis, trace fossils, unconformities, sediment gravity flows, and the origin of submarine canyons. Current research concentrates on subaqueous debris flow deposits. His primary interest focuses on inferring formative processes from the rock record. Supervision of student research projects in coal-measure geology and modern estuary sedimentology, among others, has additionally broadened his knowledge. A devotee of fieldwork, Dr. Lewis has led local and international geological excursions in the South Island of New Zealand.